MATERIALS SCIENCE AND TECHNOLOGIES

CHARACTERIZATION AND DEVELOPMENT OF NOVEL MATERIALS

MATERIALS SCIENCE AND TECHNOLOGIES

Additional books in this series can be found on Nova's website
under the Series tab.

Additional E-books in this series can be found on Nova's website
under the E-books tab.

MATERIALS SCIENCE AND TECHNOLOGIES

CHARACTERIZATION AND DEVELOPMENT OF NOVEL MATERIALS

RAFIQ ISLAM
EDITOR

Nova Science Publishers, Inc.
New York

Library of Congress Cataloging-in-Publication Data

Characterization and development of novel materials / editor, Rafiq Islam.
p. cm.
Includes index.
ISBN 978-1-61324-149-3 (hardcover)
1. Materials. I. Islam, Rafiqul, 1959-
TA403.8.C46 2011
620.1'1--dc22

2011007936

Published by Nova Science Publishers, Inc. ✝ New York

CONTENTS

PREFACE

This new book presents and discusses current research in the study of the characterization and development of novel materials. Topics discussed include the development of probabilistic simulation model for production optimization in open pit mines; lead removal from aqueous streams by a novel bioadsorbent; characterization of clinical bacteria and their control by a medicinal plant; preparation of stronium hexaferrite magnets from celestite and blue dust by mechanochemical route and the investigation of the properties of a borosilicate glass from recycled manganese crust tailings.

Chapter 1 - As the need for environmentally safe solutions to chemical engineering problems arise, natural alternatives to existing systems are being studied extensively. Past chemical engineering practices led to the generation of extremely high volumes of wastes, which pose serious long-term problems when heavy metals are present. The use of waste and naturally occurring elements as potential valuable assets helps move in the right direction, as not only is it a use for this waste, but also avoids the use of synthetic products that are often more toxic than the original toxic waste. The overall goal is to devise new solutions that are convergent over long periods of time, so that the system does not lead to irreversible damage of the earth. In this paper a novel bioadsorbent is studied. Kernels from mango stones, that are usually considered as a waste, show very high efficiency (as high as 90%) when used to remove lead contaminants. More importantly, they show delayed breakthrough, indicating high adsorptive capacity. The numbers are more impressive than conventional adsorbents, such as activated carbon or resins. Finally, mathematical models using Freundlich isotherms are developed. Using these charts the adsorption system is mathematically modeled, producing nearly accurate theoretical results to those obtained experimentally. This would help scale up the laboratory results to design a pilot plant. All design criteria are investigated and reported in this paper.

Chapter 2 - This paper presents the use of particle swarm optimization to maximize the clean coal yield of a coal preparation plant that typically has multiple cleaning circuits that produce the same product quality so that the blend of clean coal meets the targeted product quality constraints. Particle swarm is used for the yield optimization while satisfying multiple product quality constraints. The results show a 2.73% increase in the yield can be achieved leading to additional revenue of $5,460,000 per annum for a plant producing 10 million tons of clean coal per year without significantly adding to the implementation/operation cost.

Chapter 3 - Large mechanized open pit operations can become more efficient and competitive without additional costs by improving the productivity of the existing machinery.

This can be achieved by ensuring higher availability and utilization of the equipment The present paper deals with truck dispatch system using a variety of strategies to enhance the production capacity of a given set of equipment (shovels and trucks) by way of continuous dispatching of trucks. A discrete event probabilistic simulation program has been developed to demonstrate the production potential of an operating mine. It can accommodate any number of shovels and trucks of different capacities. The characteristic features of each truck and shovel can be individually specified to reflect the reality of actual working conditions. The model was tested on a hypothetical mine and a real case to verify the logic used in the program development. Subsequently, the model was implemented in two real mines to investigate the production potential. It was observed that substantial scope exists for productivity improvement when the mines were operating on a flexible mode.

Chapter 4 - A field experiment was conducted on Senna (cv – Rabi 20/105) during 2004 to 2005 at Kalyani, Nadia, west Bengal in split plot design in replicate on sandy clay loam soil with eight dates of sowing and two spacing (15 cm × 30 cm and 45 cm × 30 cm), to study their effects on the growth and growth attributes and yield of Senna. The date of sowing and spacing significantly influenced the growth attributes i.e. plant height, dry matter production, crop growth rate, Leaf area index.

Chapter 5 - In the present study celestite ($SrSO_4$), the natural ore of strontium from Trichirapalli district in Tamil Nadu (India) containing impurities like Fe, Ba, Si etc. was chemically treated with HCl/HNO_3 to reduce impurity level. After initial acid purification the mass was treated with charcoals of different grades to obtain SrS from celestite. This was finally converted to $SrCO_3$ by treating SrS with Na_2CO_3 solution. In another effort the leached celestite was directly heated with Na_2CO_3 solution to obtain $SrCO_3$. The percentage conversion of $SrSO_4$ to SrS has been studied with respect to time, temperature, mole ratio etc. The study includes X-ray diffraction study of the raw material celestite, reduced SrS and produced $SrCO_3$ for the identification of phases. The SEM study was also conducted for celestite and strontium carbonate for the study of its particle morphology. The details of conversion and the obtained results have been described in this paper. Focus has been also made to optimize the process to prepare $SrCO_3$ (ferrite grade) with maximum purity up to 99.0% $SrCO_3$. The present study also gives a comparative study of the two processes involved in the preparation of $SrCO_3$ in terms of process easiness, economy, percentage yield etc.

Chapter 6 - The effectiveness of a medicinal plant (*Cassia sophera* L.) for controlling the growth and survival of microorganisms isolated from stool samples collected from Kushtia General Hospital, Bangladesh was studied. Biochemical tests identified some bacteria such as *Staphylococcus aureus*, *Enterococcus species*, *Pseudomonas aeruginosa, Escherichia coli* and *Klebsiella pneumoniae*. The antibacterial activity of various organic solvent (petroleum ether, chloroform and ethanol) extracts of *C. sophera* leaves was determined *in vitro* using agar diffusion method and MIC against the isolated bacteria. The extracts (300 $\mu g\ disc^{-1}$) displayed potential of antibacterial activity against two Gram-positive bacteria (*S. aureus* and *Enterococcus* sp.) and two Gram-negative bacteria (*Klebsiella pneumonia* and *E. coli*). The zones of inhibition of different organic solvent extracts against the tested bacteria were found in the range of 12.1 ± 0.5 to 21.3 ± 1.1 mm and the MIC values were recorded between 62.5 and 500 $\mu g\ ml^{-1}$. The antibacterial activity was compared with commercial antibiotics such as streptomycin and tetracycline. In most of the cases, the extracts (chloroform and ethanol) exhibited similar or higher antibacterial activity than standard drug streptomycin. The results of this study suggest that the extracts of *Cassia sophera* L. can be a source of natural antimicrobial agents with potential applications in pharmaceutical industries to control pathogenic bacteria causing severe illness in humans.

Chapter 7 - Today's civilization is based on the technologies that are tangible in nature. However, tangibles relate to short-term and very limited in space whereas the intangibles relate to either long-term or to other elements of the current time frame. Therefore, a focus on tangible will continue to obscure long-term consequences. The long-term consequences will not be uncovered until intangible properties are properly analyzed and included. This study has emphasized on intangibles due to its inherent importance and showed how tangibles should link with intangibles. This has opened up the understanding of the relationship between intangible to tangible scales; from microscopic to macroscopic properties. This study also develops a guideline for sustainability using mass and energy balance which is an extended work of sustainability development by Khan and Islam, 2007a. True Sustainability Technological Development and Natural Resources Management, Nova Science Publishers, NY, USA, 381p.). The zero-waste concept is inherently related to true sustainability. In this study, a detailed analysis of different features of sustainability is presented in order to understand the importance of using the concept of sustainability in every technology development model. It was identified that for a process to be sustainable, it must have the source as well as the pathway be natural. Observation and understanding of nature, mass and energy balance and the relationship, activities etc. were the research tools to explain the importance of intangibles in human life and technology development. The current study takes a significant step forward developing the link between 'hard science' and the science of intangibles.

Chapter 8 - In the present investigation celestite (natural ore of strontium) and blue dust (iron ore fines) have been used for the preparation of strontium hexaferrite powder. The mechanical alloying process has been adopted to prepare strontium hexaferrite powder. The celestite after chemical upradation and physically upgraded blue dust alongwith sodium carbonate was taken for the preparation of strontium hexaferrite in this experiment. The high-energy planetary ball mill with tungsten carbide jar and ball was used to prepare strontium hexaferrite powder. A long time of ball milling for different duration has led to displacement solid-state reaction. At the end of each experiment the product was washed thoroughly and dried. The X-ray diffaction study after annealing shows the development of single-phase strontium hexaferrite after 40 hrs. of milling. The resultant powder was compacted under magnetic field and sintered to prepare the magnet after annealing the ferrite powder. The magnetic properties were measured by Pulse magneto meter. The moderate value of coercivity, remanence and energy product were observed in this sintered magnet. The work illustrates the feasibility to prepare strontium hexaferrite magnetic powders directly from natural ores which can reduce the total cost of production as compared to conventional method.

Chapter 9 - A simple high-performance thin-layer chromatographic (HPTLC) method has been developed for the simultaneous determination of the pharmacologically important two hydroxy anthracene glycosides *viz.* sennoside A and sennoside B in *Cassia angustifolia*. The assay combines the separation and quantification of the analytes on silica gel 60 GF_{254} HPTLC plates with visualisation under UV and scanning at 350 nm. Using this technique, the alkaloidal content of different parts of the title plant have been determined.

Chapter 10 - Manganese crust tailings vitrification with boric acid as flux and silica as an additive at 1160C was aimed at recycling waste tailings into borosilicate glass materials.

Glass products characterization and testing was conducted using TGA, DTA SEM/EDX and optical microscope and revealed acceptable physical and chemical characteristics. Glass 5-15 exhibited a density of 2.3 g/m^3, H_v of 4.8 GPa, K_c of 0.66 MPa m$^{1/2}$, E value of 92, TEC of 2.89 in x 10^{-6} and acceptable chemical resistance. The glass was able to be drawn into fibers of 200 to 400 micron diameter, which suggests its potential use in producing high-end marketable products and indicates the possibility of future research into additional applications.

Chapter 11 - The present investigation was conducted to find out the utilization of effective bacteria for solid kitchen waste degradation, and subsequent uses of degraded substances as organic manure or compost. Ten garbage samples were collected from different potential habitat of Kushtia and Jhenaidah regions of Bangladesh for isolating the effective garbage degrading bacteria. Three bacterial sample strains named as NAIUL, BCDump and BCSS (2) were isolated and cultured on Nutrient Agar and Czapek-Dox-Agar media. Optimum cultural conditions (pH of preferred medium, temperature, incubation period) of the bacteria were studied before massive multiplication. Characterizations of the strains were also studied by visual observation of colony, microscopic observation and biochemical tests identified the specific bacteria namely *Xanthomonas* spp., *Bacillus* sp., and *Pseudomonas* sp. from NAIUL, BCDump and BSS (2) strains, respectively. Gram staining reaction and motility test were studied by microscopic observation. All the strains were Gram positive and motile. Biochemical tests ensured their physiological characteristics. To select the most effective strain for decomposition of waste, autoclave sterilized kitchen garbage was used and inoculated with bacterial suspension and culture pellet. Changes of color, odor, weight loss, volume loss, temperature and pH of decomposing garbage were noted. The highest decomposition activity in term of weight loss and volume loss was shown by suspension of BCDump sample strain. To observe the effect of various additives on decomposition, various concentrations (5 %, 10 % and 15 %) of sucrose and molasses solution were used along with bacterial suspension. Among different sucrose treatments, BCDump sample strain along with 15 % sucrose solution was found to be the most effective. As the sucrose solution treatment showed better performance as additive, various concentrations of molasses solution (5 %, 10 % and 15 %) were also tested along with bacterial suspension as an alternative and cheap source of carbohydrates. It was observed that 15% molasses treatment showed the best performance over the other treatments using BCDump sample strain. Decomposed garbage was used as compost to observe their effects on biomass production (fresh weight basis) of potatoes (*Solanum tuberosum*). Garbage decomposed by all strains was used as organic manure. From the present investigation, it can be concluded that useful bacteria might be isolated from the surrounding environment for friendly bioconversion of solid organic waste. To protect human health and natural environment from waste problems, the developed method of decomposition of waste as shown in this study is found to be very effective and useful.

Chapter 12 - This paper has discussed the oil potential of Karanj (Pongamia pinnata) as source of bio-diesel, for cultivation in Bangladesh and other potential uses. This paper has also outlined the oil extraction method from the seed which is s chemical solvent method known as cold percolation method by using the CCl4 and this method can extract more than 95% oil. Physical properties of the Karanj seed are also discussed and the bulk density of the seed is 0.56g/cc. Karanj seed contains oil about 32%. The specific gravity of Karanj oil varies from 0.925 to 0.940 at 200 C which is slightly higher than the sp. gravity of fossil diesel.

Therefore, it should be adjusted for its large scale application in engine running. The high viscosity of Karanj oil interferes with injection process and leads to poor fuel atomization. The injection system may be slightly modified for the application of Karanj oil as fuel for diesel engine or the high viscosity has to be overcome by using methyl ester with Karanj oil. The transformation of Karanj oil to its methyl ester reduces molecular weight to one third, reduces viscosity to one eighth and increases the diesel index. The flashpoint of the oil is 134 °C which explain that the Karanj oil could stand for a long time. So, Karanj cultivation in Bangladesh could be a practical solution for the rural people in order to minimize the energy shortage.

Chapter 13 - Conventional rotary drilling has been the main technique used for drilling in the oil and gas industry. However, this method has showed its limitations regarding the depth of the wells drilled in addition to the use of toxic drilling fluids. This paper proposes an alternative method that is waterjet drilling. This drilling technique is environmentally friendly, fast, and feasible. Based on laboratory experiments, empirical models have been developed to describe the jet-rock interaction simulated by using beeswax samples. The resourceful and conclusive results show that the depth and rate of penetration increase with time, temperature, and thermal exposure.

Chapter 14 - The use of supported ionic liquid membranes, which consist of a thin microporous support whose pores are filled with a liquid, have been extensively studied during the last decade. Supported ionic liquid membranes have being recognized as an effective method for the selective separation of organic compounds and mixed gases. The unique properties of ionic liquids such as their negligible vapour pressure, their greater viscosity and the possibility of minimizing their solubility in the surrounding phases by adequate selection of the cation and anion could overcome the instability associated with supported liquid membranes based on organic solvents. Furthermore, the possibility of design suitable ionic liquids for specific separation problems has also opened up new fields of application of supported ionic liquid membranes. In this review an overview is given of recent advances in supported membranes based on ionic liquids, including issues such as methods of preparation, physical–chemical characterization, stability and long–term performance and fields of application of supported ionic liquid membranes.

Chapter 15 - The cracking stress and bending moment of ferrocement elements in flexure are the significant parameters in the design of structures made of ferrocement elements. This paper presents an experimental investigation on ferrocement elements uniformly reinforced by four different meshes, such as, geogrid mesh, chicken mesh, square mesh and hybrid mesh. The cracking stress and bending moment have been studied by testing the specimens under flexural loadings. By analyzing experimental data, the equations for the first-crack-stress and the ultimate-moment have been proposed in this paper. Results obtained by the proposed equations have been compared to those reported in the literature. It was revealed that the proposed equations, irrespective of their simplicity; give good agreement to those previously published.

Versions of these chapters were also published in *Journal of Characterization and Development of Novel Materials*, Volume 1, Numbers 1-4, edited by Rafiq Islam, published by Nova Science Publishers, Inc. They were submitted for appropriate modifications in an effort to encourage wider dissemination of research.

In: Characterization and Development of Novel Materials
Editor: Rafiq Islam
ISBN: 978-1-61324-149-3

Chapter 1

LEAD REMOVAL FROM AQUEOUS STREAMS BY A NOVEL BIOADSORBENT

Moniruzzaman Khan and M. Rafiqul Islam*
Department of Civil and Resource Engineering, Dalhousie University
Halifax, Canada

ABSTRACT

As the need for environmentally safe solutions to chemical engineering problems arise, natural alternatives to existing systems are being studied extensively. Past chemical engineering practices led to the generation of extremely high volumes of wastes, which pose serious long-term problems when heavy metals are present. The use of waste and naturally occurring elements as potential valuable assets helps move in the right direction, as not only is it a use for this waste, but also avoids the use of synthetic products that are often more toxic than the original toxic waste. The overall goal is to devise new solutions that are convergent over long periods of time, so that the system does not lead to irreversible damage of the earth. In this paper a novel bioadsorbent is studied. Kernels from mango stones, that are usually considered as a waste, show very high efficiency (as high as 90%) when used to remove lead contaminants. More importantly, they show delayed breakthrough, indicating high adsorptive capacity. The numbers are more impressive than conventional adsorbents, such as activated carbon or resins. Finally, mathematical models using Freundlich isotherms are developed. Using these charts the adsorption system is mathematically modeled, producing nearly accurate theoretical results to those obtained experimentally. This would help scale up the laboratory results to design a pilot plant. All design criteria are investigated and reported in this paper.

Keywords: bioadsorbent; mango kernel; heavy metal; Freundlich isotherm.

* Corresponding author: E-mail: rafiqul.islam@dal.ca.

1. Introduction

The ecological balance is being disrupted continuously due to activities of industrialization and urbanization. These activities have led to the alarming presence of heavy metals such as lead, copper, zinc and cadmium in waste streams, sewage and rivers. Heavy metals are classified as persistent environmental toxic substances because they can not be rendered harmless by chemical or biological remediation processes (Devi et al., 1995). Among the heavy metals, the toxicity of lead has been known since ancient times because of its wide use both in industry and in household products for centuries. Recently, much research has focused on the treatment of effluents containing lead. Several adsorbents have been found that adsorb lead from the contaminated water. However, most existing treatment technologies based on synthetic polymeric resin or other synthetic products as adsorbents again pose serious environmental problems due to secondary toxicity which is more toxic than original one. That is why recent research attention has focused on natural organic, living or non-living biosorbents, particularly low-cost and waste materials (Kim, 2003; Mustafiz et al., 2003). The true benefit of any system is achievable if a natural pathway is followed (Khan et al, 2007; Khan and Islam, 2007).

Some natural organic materials such as Nonliving biomass of Sargassum, Lignocellulosic materials and wastes such as peanut skin, cotton, onion skin, fish scale, wood saw, rice hulls, maize stalks, barks, jute fibers, bagasse, rice straw, corncobs, husks and palm stone kernels etc not only have excellent adsorbability of heavy metals, but also have biocompatibility, biodegradability, and nontoxicity (Gin et al., 2002; Basu, 2005). The aim of this paper is to study kernels from mango stone for the removal of lead ions. Mango stones are known as the wastes after consuming mangoes. A large amount of mango stones can be obtained economically from food processing plants as process wastes.

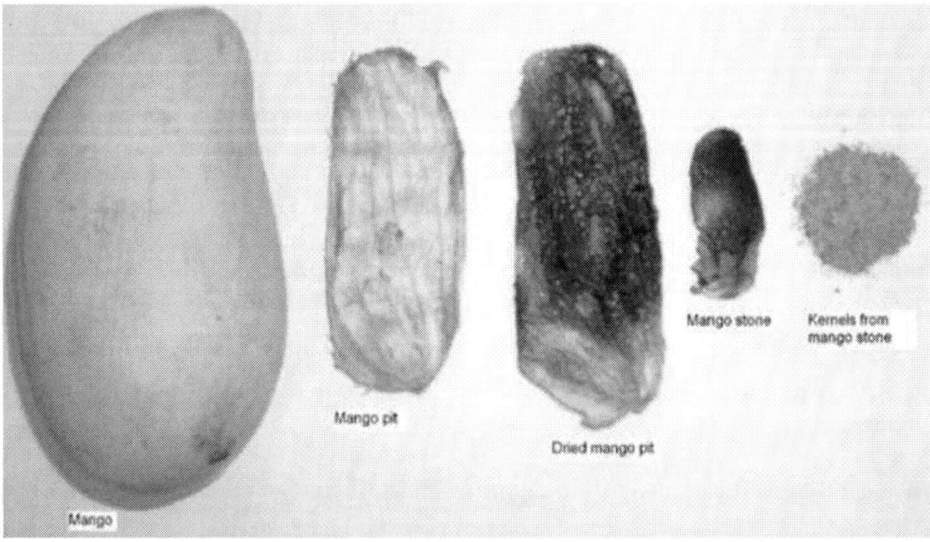

Figure 1. Kernels from mango.

2. Materials and Method

2.1. Natural Adsorbent

Mango kernels were used as natural absorbent for heavy metal absorption. Mango pits were collected from the mangoes. Mangoes were collected from Pete's Frootique, Nova Scotia, Canada. Mango stones are naturally covered by mango-pits. Mango-pits were dried to reduce moisture to nearly 10% and the mango stones were separated from mango-pit. The mango stones were ground and sieved to obtain 60 to 100 US mesh size of kernels (Figure 1).

2.2. Adsorption Experiment

Sorption experiments were carried out in batches in 25 ml test tube containing 20 ml $Pb(NO_3)_2$ solutions of different lead concentrations that varied from 1 ppm to 1000 ppm. Sufficient stock solution that had a lead concentration of 1000 ppm was prepared in the mining laboratory of Dalhousie university from lead nitrate and deionized water. Later on, various concentrations of lead solutions were prepared from the stock solution by adding a calculated amount of deionized water. Dried kernels from mango stone (0.5g) were added to each test tube which was then sealed with a plastic cap to minimize evaporation and shaken vigorously to have a good mixing. The experiments were carried out at room temperature of 23° C (± 1° C). The pH of some samples was varied by using buffer solutions (HCL or NaOH) for getting acidic or basic solution.

2.3. Analytical /Chemical Analysis

The pH of the samples was measured by pH meter and lead concentrations were determined by an Atomic Absorption (AA) technique after 48 hours of preparation. No further distinct change in lead concentration was noticed after that time. The concentrations of lead ions after 48 hours were taken as equilibrium concentrations.

3. Results and Discussion

The adsorption of Pb^{2+} onto the surface of mango-stones is the aim of this study. One of the most interesting methods for the removal of heavy metals from various effluents involves the adsorption capacity of natural organic or inorganic materials which are particularly abundant and inexpensive (Kumar and Das, 2000). The adsorption capacity of any material might be determined by the use of an adsorption isotherm. Several isotherms have been developed so far, however, Freundlich and Langmuir isotherms are the most popular and widely used isotherms in wastewater treatment applications. The number of adsorbent sites mostly determines the amount of adsorbate that will adsorb onto the adsorbent. The number of these sites is quantified by the ratio of sorbed materials to the adsorbents as soon as equilibrium is established between adsorption and desorption (Sincero et al., 2002). The equation of Freundlich isotherms is shown below:

$$\frac{x}{m_a} = k[C]^{1/n} \quad (1)$$

where,

x = Mass of the adsorbate adsorbed onto the mass of adsorbent, m_a
m_a = Mass of adsorbent
[C] = Lead concentration in the solution in equilibrium
n and k are constants
The straight-line forms of the Freundlich isotherms is,

$$\ln \frac{x}{m_a} = \ln k + \frac{1}{n} \ln[C] \quad (2)$$

It is known that the amount of adsorbent influences the adsorption. That is why the graph, showing the relation of the initial metal concentration and the equilibrium concentration, has little significance. However, this graph gives a sense of the trend of adsorption. Figure 2 shows the trend of adsorption for a 0.5 g of adsorbent in 25 ml of lead solution.

The above figure also shows that the amount of adsorption varies with different concentrations even if the amount of adsorbent is same. This indicates that not only the amount of adsorbent but the initial concentration is also important for the adsorption process. The dynamic adsorption and desorption ultimately bring an equilibrium concentration beyond which no further adsorption takes place.

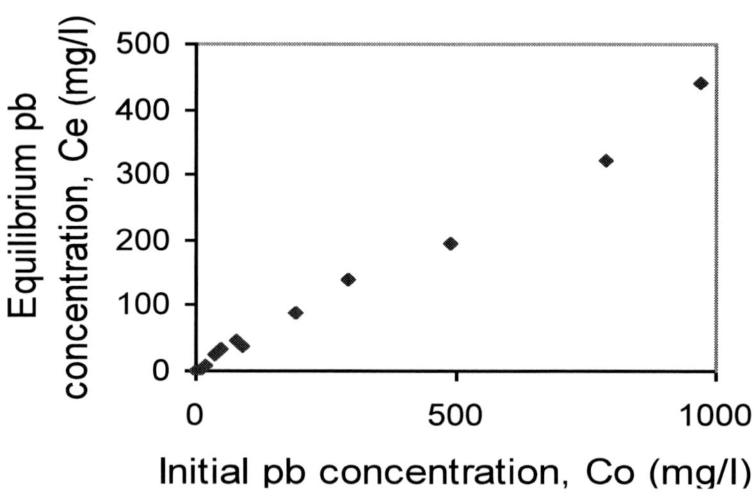

Figure 2. Equilibrium concentration (mg/l) in 20 ml lead solution for 0.5 g mango kernels.

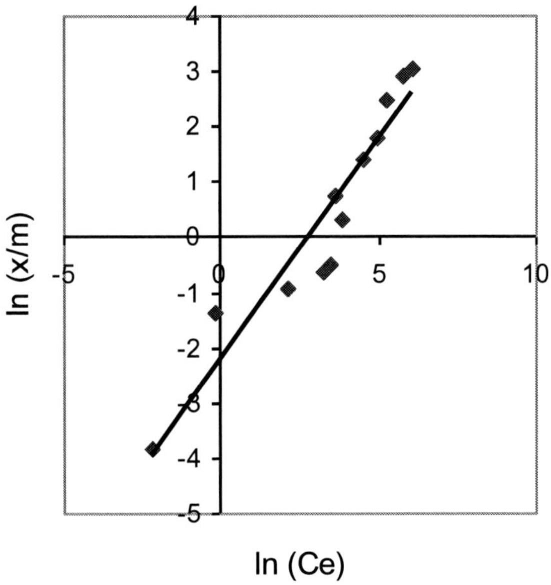

Figure 3. Freundlich isotherm.

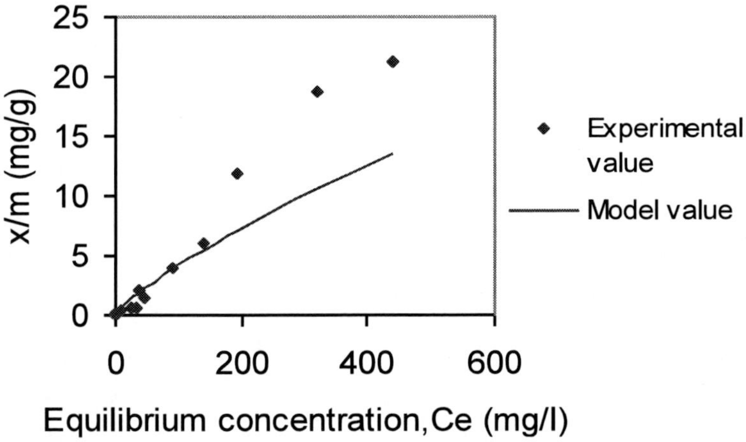

Figure 4. Comparison of Freundlich model with experimental.

The isotherms explain the adsorption phenomena. The Freundlich model was used to explain the observed phenomena. The equilibrium data were analyzed using Eq. 2. In this model, k and n are Freundlich empirical constants related to adsorption capacity and intensity respectively. Our interest is to measure lead adsorbed onto the kernels of mango at equilibrium lead concentration. As the experiments were performed at one temperature (24 °C), we have got one value for each constant. From the plot (Figure 3), we calculated the value of k and 1/n which were 0.1106 and 0.79, respectively. The value of 1/n indicated a favorable adsorption as it is less than unity. Again the value of k indicates the acceptable adsorption capacity. The Freundlich equation can be written as:

$$\frac{x}{m_a} = 0.1106\,[C]^{0.79} \qquad (3)$$

After obtaining the co-efficient of the Freundlich model, we have compared the experimental data with this model (Figure 4).

The model was well fitted with the experimental value for lower equilibrium concentration but showed a little variation in the upper concentration. The variation might be due to heterogeneity of organic materials. Figure 5 shows the heterogeneity of the surface of kernels obtained from mango stone. Freundlich model is based on empirical formulae. So, it is unexpected to fit the data perfectly with this model.

3.1. Effect of pH

In most cases, the pH of the solution has an impact on the metal uptake. To study the effect of pH, the same initial concentration (30 mg/l) of different pH solutions were prepared and one gram of kernels was added to each solution. The effect of pH on the metal uptake was observed. Three solutions were used; those were:

- Acidic solution (3.7 pH)
- Basic solution (10.8 pH)
- Natural (close to neutral) solution (6.44 pH)

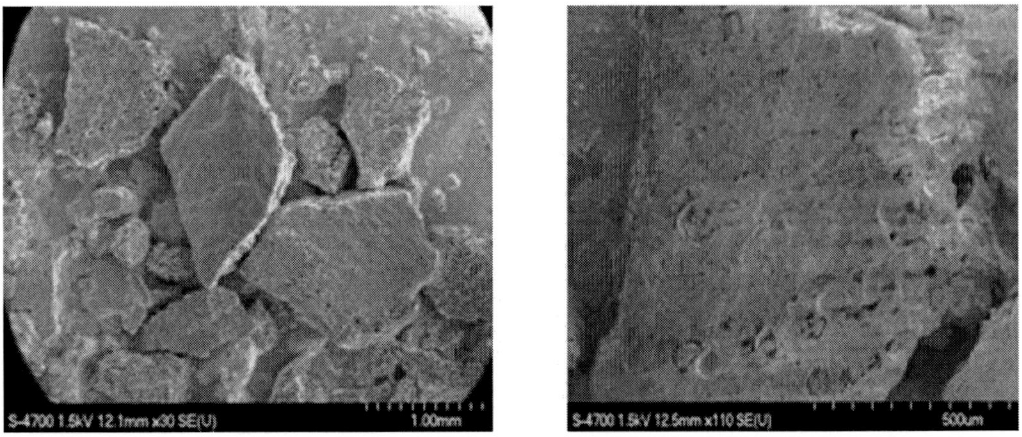

Figure 5. SEM of mango-stone kernels.

For the same initial concentration, no noticeable change in equilibrium concentration was observed (Figure 6). This reveals that the adsorption of lead onto the kernels of mango stones is independent of pH. Some times, pH accelerates the precipitation process (Zhou et al., 2004). In that case, metal accumulation on the surface of adsorbent is not considered due to adsorption alone. Precipitation also takes place. However, if the pH changes do not contribute to the change of equilibrium concentration, the metal accumulation can be considered due to adsorption alone without any contribution of precipitation. In other words, it is denoted that if the pH is changed due to foreign materials, the adsorption will not change due to acidity or alkalinity. In this experiment, the metal accumulation is considered due to adsorption alone as pH variation did not change the equilibrium concentration appreciably.

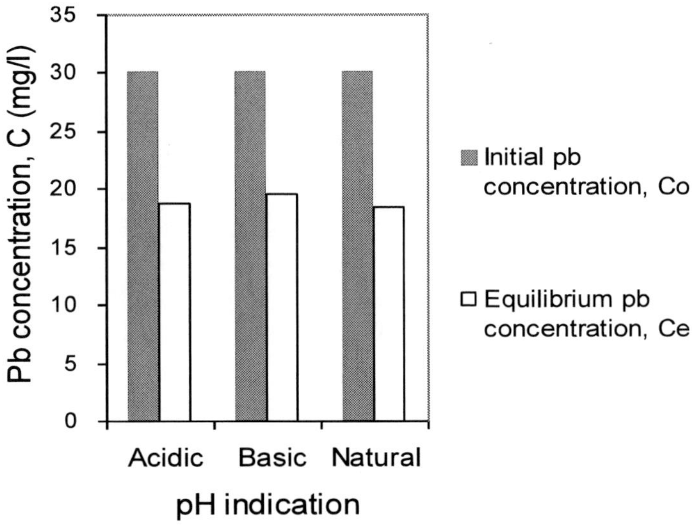

Figure 6. Effect of pH on lead adsorption.

Conclusion

This study showed that Kernels from mango stones can be used as a new adsorbent for the removal of lead. The pH does not affect the adsorption. That is why kernels from mango stones can be used for any solution with any pH. The heterogeneity of the bio-material is well known and to reach a constant value will be difficult. After use, the substrate can finally be converted into non-toxic composts and thus contribute to zero-waste concept.

Acknowledgment

The authors would like to acknowledge the contribution of the Atlantic Innovation Fund (AIF).

References

Basu, A., (2006) Experimental and Numerical Studies of a Novel Technique for Abatement of Toxic Metals from Aqueous Streams, Ph.D. Dissertation, Faculty of Engineering, Dalhousie University, Canada, 304p.

Devi, M. and Fingerman, M., (1995) Inhibition of Acetylcholinesterase Activity in the Central Nervous System of the Red Swamp Crayfish, Procambarus Clarkii, by Mercury, Cadmium, and Lead, *Bulletin of Environmental Contamination and Toxicology*, vol. 55, No. 5, pp. 746.

Gin, K.Y.H., Ying-Zhong Tanga, Y.Z. and Aziz, M.A.,(2002) Derivation and Application of a New Model for Heavy Metal Biosorption by Algae, *Water Research*, vol. 36, pp. 1313–1323.

Khan, M.I. and Islam, M.R., (2007) True Sustainability in Technological Development and Natural Resources Management. Nova Science Publishers, New York, USA, 381 p.

Khan, M.M, Prior, D. and Islam, M.R., (2007) Zero-Waste Living with Inherently Sustainable Technologies, *Journal of Nature Science and Sustainable Technology*, vol. 1, No. 3.

Kim, D. S., (2003) The removal by crab shell of mixed heavy metal ions in aqueous solution, *Bioresource Technology*, vol. 87, No. 3, pp. 355-357.

Kumar, N. and Das D., (2000) Enhancement of Hydrogen Production by Enterobacter cloacae IIT-BT 08, *Process Biochem*, vol. 35, No.6, pp. 589–93.

Mustafiz, S., Rahman, M.S., Kelly, D., Tango, M., and Islam, M.R., (2003) The Application of the Fish Scales in Removing Heavy Metals from Energy-Produced Waster Streams: The Role of Microbes, *Energy Source*, vol. 24, pp. 905-916.

Sincero, A. P. and Sicero, G. A., (2002) Physical-Chemical Treatment of Water and Wastewater, IWA Publishing, London, UK.

Zhou, D., Zhang, L., Zhou, J. and Guo, S., (2004) Development of a Fixed-Bed Column with Cellulose/Chitin Beads to Remove Heavy-Metal Ions, *Journal of Applied Polymer Science,* vol. 94, pp. 684–691.

In: Characterization and Development of Novel Materials ISBN: 978-1-61324-149-3
Editor: Rafiq Islam

Chapter 2

Particle Swarm Optimization Approach for Maximizing the Yield of a Coal Preparation Plant

Vishal Gupta, Hussain Unjawala, Ajay Mahajan*♦ *and Manoj Mohanty
College of Engineering
Southern Illinois University at Carbondale
Carbondale, IL, US

Abstract

This paper presents the use of particle swarm optimization to maximize the clean coal yield of a coal preparation plant that typically has multiple cleaning circuits that produce the same product quality so that the blend of clean coal meets the targeted product quality constraints. Particle swarm is used for the yield optimization while satisfying multiple product quality constraints. The results show a 2.73% increase in the yield can be achieved leading to additional revenue of $5,460,000 per annum for a plant producing 10 million tons of clean coal per year without significantly adding to the implementation/operation cost.

Keywords: Coal, optimization, particle swarming, coal plant optimization.

1. Introduction

A modern coal preparation plant flowsheet integrates three or four cleaning circuits that cleans the coarser (+ 6 mm), intermediate (6 x 0.6 mm) and finer (- 0.6 mm) size fractions. Figure 1 shows a block diagram of a coal processing plant that runs with a four unit

♦ Corresponding author. Email: mahajan@engr.siu.edu, Tel: 618-453-7007, Fax: 618-453-7658.

operations which are categorized as heavy medium vessel (HMV), heavy medium cyclone (HMC), spiral and froth-flotation. The feed is screened and classified into different size fractions to be treated in respective circuits. The heavy medium vessel cleans the coarsest (152 x 16 mm) size fraction while heavy medium cyclone treats the intermediate (16 x 1 mm) size particles.

The fine (1 x 0.15 mm) and ultra-fine (0.15 x 0.045 mm) size fractions are cleaned in spiral and flotation processes, respectively. The size fraction below 0.045 mm is discarded as slimes due to its high ash content. Each circuit is operated at specified operating conditions to produce the same product quality as the targeted plant product quality. The key operating conditions may be listed as medium density for HMV and HMC splitter position for coal spirals, and residence time for flotation process. Although, the targeted overall plant product quality is satisfied by this approach, the plant yield is not maximized. The output of normal plant operation is considered as baseline for further comparison with the different optimization models used in the study.

Numerous studies have been conducted in the past to optimize the clean coal yield (defined as percentage of plant feed reporting to the plant product stream) of a coal preparation plant. A graphical approach was suggested by Sarkar *et al.* (1960) for maximization of yield of composite clean coal at desired ash content. It was suggested that cleaning of coarser coal at higher ash content and finer coal at relatively lower ash content gives the maximum yield while satisfying the product ash constraint. Additional graphical approaches were suggested by other investigators (Salama, 1991, Lyman, 1993 and Salama, 1998) to optimize the yield of a coal preparation plant on the basis of product ash. The yield-ash curves were plotted for each circuit. The points of equal slope were located on these curves and the one that met the product ash constraint was considered as the optimum yield of the plant. Salama (1986) suggested a numerical approach to maximize plant yield. A Lagrangian approach was used for a constrained nonlinear optimization process.

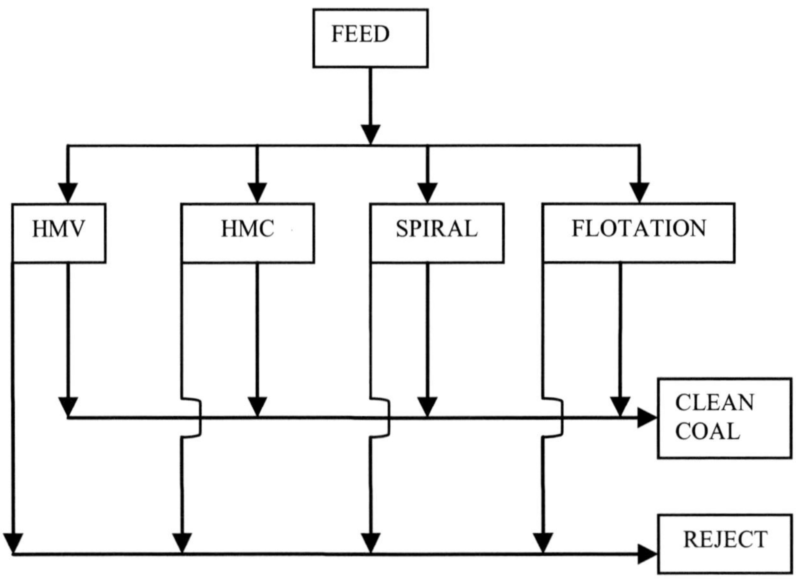

Figure 1. A simplified block diagram of a four circuit coal processing plant.

Table 1. Operating conditions maintained during plant-testing carried out during this investigation

Test	Spiral Splitter Position		HMV Medium Density	HMC Medium Density
	A (inch)	B (inch)		
1	6.5	0.25	1.32	1.32
2	6.5	0.5	1.36	1.36
3	6.5	0.75	1.40	1.40
4	6.5	1	1.44	1.44
5	6.5	1.25	1.48	1.48

The product ash was considered as the only constraint for the yield optimization process. The necessary condition was determined to maximize the yield by equalizing the incremental product ash for each circuit. Honaker *et al.* (1997) and Luttrell *et al.* (2004) also arrived at the same conclusion that the equal incremental ash approach gives the maximum yield at target ash content. Gupta *et al.* (2004 and 2005) developed a model using genetic algorithms and concluded that genetic algorithm serves as better alternative for coal preparation plant yield optimization in comparison to the equalization of incremental product ash approach when it is required to satisfy multiple product quality constraints.

The present study describes the application of an evolving evolutionary algorithm, known as particle swarm optimization (PSO) for solving the coal plant yield maximization problem while satisfying multiple product quality constraints. The PSO is a stochastic optimization technique that can be compared to the behavior of a flock of birds or the sociological behavior of a group of people (Bergh and Engelbrecht, 2004). This technique has been used to solve a range of optimization problems, including neural network training and function minimization. Kennedy and Eberhart (1995) used particle swarming for a neural net application and concluded that it could train neural network weights as effectively as the usual error back propagation method. The particle swarm optimization has also been used to train a neural network to classify the Fisher Iris Data Set (Fisher, 1936). Over a series of ten training sessions, the trained weights found by particle swarms sometimes generalize from a training set to a test set better than solutions found by gradient descent. Pomeroy (2003) discusses the application of particle swarm optimization in optimizing a function whose value depends on four dimensional position vectors.

This paper considers the yield optimization of the coal preparation plant based on two product quality constraints, i.e., product ash assay and sulfur assay. Two approaches, i.e., traditional incremental ash based approach and PSO, were investigated to maximize plant yield while satisfying the target product ash and sulfur assays. The optimal yields obtained from both approaches have been compared.

2. Data Analysis

The coal preparation plant, studied in this investigation, receives coal from two different sections of a mine blended at a ratio of 75% from Section-1 and 25% from Section-2. The operating conditions for each unit operations (HMV, HMC, Spiral and Flotation Cell) were varied in succession to conduct a series of experiments and generate characteristic partition curves for each process. The medium specific gravities were varied over a range of 1.32 to 1.48 for the HMV and HMC as indicated in Table 1.

For Spiral, the product splitter (B) position was varied over 5 individual locations. The partition data for each unit operation were generated by plotting percentage of feed corresponding to a specific density class that reported to the product stream. Model equations were developed to best fit the normalized (reduced partition) partition data generated for each density based process. The kinetic rate equation was developed for the flotation process in the lab (using the same coal) since the flotation tests conducted in the plant failed to produce meaningful results. Model equation for each process was simulated to generate a huge amount of data for the process yield and corresponding product ash and sulfur assays for small incremental change in operating condition, i.e., for each 0.01 unit change in the separation density (D_{50}) for density based process and each 10 second increase in the residence time for the froth flotation process. These data were stored in a database to be used for the optimization model.

3. Particle Swarming Optimization Model

The particle swarm algorithm works by searching iteratively in a region that is defined by each particle's best previous success, the best previous success of any of its neighbors, the particle current position, and its previous velocity (Mendes *et al.*, 2004). A particle searches through its neighbors in order to identify the one with the best result so far and uses information from that one source to bias its search in a promising direction.

A simplified block diagram of a modern four circuit coal preparation plant investigated during this study is shown in Figure 1. The analogy of finding an optimal yield of the plant is similar to a flock of birds searching for the roost area. The flock of birds has a common behavior in which they fly in a group in search for a roost area. All other birds follow their neighbor. The neighbor follows their neighbors and so on. So, if any single bird finds the roost area, all the birds eventually group around that area. In the coal plant optimization problem, the yield value from each circuit at different operating conditions can be thought of single dimension and so on. An initial population of particles was created as a function of yield value of each circuit which can be given as:

$$\text{Particle population} = [x\ y\ z\ w] \quad (1)$$

where,
x = yield values of heavy medium vessel process at different specific gravity
y = yield values of heavy medium cyclone process at different specific gravity
z = yield values of spiral process at different specific gravity
w = yield values of flotation process at different residence time

The yield values of respective circuit as represented by x, y, z and w were generated randomly but within realistic limits as determined from the experimental results. The position of a particle is defined in terms of four dimensional x, y, z and w vector in a solution space. A set of random numbers were generated to create neighbors for each particle. For an initial population of forty particles, five neighbors were defined for each particle. These five neighbors were chosen among the population of particles excluding the particle itself.

$$\text{Neighbor particle} = [n_1\ n_2\ n_3\ n_4\ n_5] \tag{2}$$

where, n_1,n_2,n_3,n_4 and n_5 = neighbors

Also, an initial velocity vector was generated randomly for each particle. This was done so that the particles have a finite initial velocity in the solution space. This particle velocity vector is given below:

$$\text{Particle velocity} = [v_x\ v_y\ v_z\ v_w] \tag{3}$$

The velocity vector is a four dimensional vector, for the movement in each dimension represented for each circuit. For example, v_x is a velocity with which a particle changes its position in the x- dimension and so on, i.e., change in yield value for the heavy medium vessel process.

An objective function is defined to evaluate the fitness value of each particle. The objective function can be given as:

$$\text{Objective function} = f(Y, \Delta A, \Delta S) \tag{4a}$$

$$= w_y Y + w_a \Delta A + w_s \Delta S \tag{4b}$$

where,

Y = Overall yield value of the plant, i.e., weighted mean of the yield value of each circuit
= g(x, y, z, w)
ΔA = Difference in actual ash content and targeted ash content
= targeted ash content – actual ash content
ΔS = Difference in actual sulfur content and targeted sulfur content
= targeted sulfur content – actual sulfur content

w_y, w_a, w_s = weights for each cost component

The multiple product quality constraints are incorporated in the fitness function. The weights w_y, w_a and w_s are very sensitive to the specific type of constraints and need to be optimized with the change in product quality constraints.

The particle fitness is evaluated with the above objective function. The particle remembers its best position in the successive generation known as *pbest*. Also, the particle remembers the best fitness value of its neighbors it has achieved so far known as *gbest*. The fitness value of the particle (*pbest*) thus determined is compared with the best fitness value of the particles in its neighborhood (*gbest*). The best particle is chosen as the particle with the maximum fitness value. If the best particle is some particle in the neighborhood then the particle will orient itself toward the best particles by changing its position using the velocity vector. The particle updates its velocity and also changes its position as follows:

$$v[v'_x\ v'_y\ v'_z\ v'_w] = v[v_x\ v_y\ v_z\ v_w] + c1 * rand(0,1) * (pbest[x_p\ y_p\ z_p\ w_p] - present[x'\ y'\ z'\ w']) + c2 * rand(0,1) * (gbest[x_g\ y_g\ z_g\ w_g] - present[x'\ y'\ z'\ w']) \quad (5)$$

$$present\ [x''\ y''\ z''\ w''] = present\ [x'\ y'\ z'\ w'] + velocity\ [v'_x\ v'_y\ v'_z\ v'_w] \quad (6)$$

where,

present $[x'\ y'\ z'\ w']$ = Particle with current position
present $[x''\ y''\ z''\ w'']$ = Particle with updated position
velocity $[v_x\ v_y\ v_z\ v_w]$ = Particle current velocity vector
velocity $[v'_x\ v'_y\ v'_z\ v'_w]$ = Particle updated velocity vector
rand(0,1) = random number between 0 and 1
pbest$[x_p\ y_p\ z_p\ w_p]$ = Particle best position in successive generations
gbest$[x_g\ y_g\ z_g\ w_g]$ = Best neighbor particle
c1 and c2 = Stochastic weight factor

This process is repeated over the generations. The process stops when there is no change in the fitness value of the particles in consecutive generations.

4. Results and Discussion

The optimal yield of the plant determined from equal incremental product ash approach and particle swarm optimization over the baseline is presented in Table 2. It was revealed that plant produces 52% yield for the targeted plant constraints of 6.75% product ash and 1.3% product sulfur.

The same set of data was used for equal incremental ash analysis and particle swarming based models. The two models were simulated to determine the maximum yield for satisfying the plant product quality constraints of 6.75% ash and 1.3% sulfur. It can be observed from the simulation data presented in Table 1 that 2.73% higher yield can be generated from both equal incremental ash quality approach and particle swarm optimization. The results presented in Table 2 also recommend the necessary operating condition for each unit operation to achieve the optimal plant yield. Whereas the results listed for HMV and HMC are self explanatory, the set of operating condition and result listed for spiral appear to have some anomaly. Moving the product splitter (B) more inward helped increase the yield from 57.13% (for existing plant) to 67.16% (for PSO Approach) and 67.09% (for Incremental Ash Approach). However, the product ash content obtained from spiral decreased to 4.73% and 4.72%, respectively, instead of increasing from 5.22% obtained for the existing plant with this change in operating condition. It may be noted that the input data used for generating the simulated results from PSO and Incremental Ash optimization approaches were based on the size-by-size washability analysis conducted on a composite spiral feed sample collected over the entire period of spiral test program (almost 8 hours). On the other hand, the spiral product ash content listed for the existing plant was obtained from the product sample collected over only ½ hour period that Test 2 (listed in Table 1) was conducted. The authors believe that variation in plant feed characteristics may have caused the resulting anomaly.

The increase in the plant yield may have significant impact on the overall profitability of the plant. In terms of monetary value, even a 0.5% incremental in the yield of the plant by merely changing the operating settings of each circuit may contribute significantly to the profitability of the plant. For example, assuming the plant capacity of 10 million tons of clean coal per annum, and $20 per ton the selling price of the coal, an increment of 2.73% high yield may generate an additional revenue of $5,460,000 per annum.

Table 2. Comparative plant optimization results obtained from this investigation

Comparative Optimal Results					
	HMV	HMC	Spiral	Flotation	Overall
Feed Characteristics					
Ash (%)	37.13	50.23	33.37	49.66	42.61
Sulfur (%)	0.94	1.12	2.06	0.74	1.17
Product Characteristics					
Operating	D50	D50	Splitter Position	Residence	
Conditions	1.50	1.44	A = 6.5", B = 0.5"	Time (168 s)	
Yield (%)	63.49	39.60	57.13	55.01	52.00
Ash (%)	6.81	6.67	5.22	10.86	6.75
Sulfur (%)	1.11	1.27	1.11	0.90	1.18
PSO Approach					
Product Characteristics					
Operating	D50	D50	Splitter Position	Residence	
Conditions	1.69	1.61	A = 6.5", B = 1.40"	Time (93.89 s)	
Yield (%)	63.87	42.78	67.16	47.36	54.73
Ash (%)	7.03	7.15	4.73	9.95	6.75
Sulfur (%)	1.07	1.33	1.04	0.90	1.15
Incremental Ash Approach					
Product Characteristics					
Operating	D50	D50	Splitter Position	Residence	
Conditions	1.69	1.61	A = 6.5", B = 1.46"	Time (95.33 s)	
Yield (%)	63.89	42.76	67.09	47.52	54.73
Ash (%)	7.04	7.15	4.72	10.01	6.75
Sulfur (%)	1.07	1.33	1.04	0.90	1.15

CONCLUSION

Particle swarm optimization is a new generation of evolutionary algorithms and is used for the first time for optimizing the clean coal yield obtained from a coal preparation plant while simultaneously satisfying multiple product quality constraints. It is a non-linear way of optimizing the yield of a coal preparation plant and is suitable for any set of discrete data. Both, equal incremental ash quality approach and particle swarm optimization generates an additional yield of 2.73% over the normal operation of the plant contributing $5,460,000 per annum towards the profitability of a plant producing 10 million tons of clean coal per year. The particle swarming based optimization can be generalized for any number of product quality constraints. The product quality constraints just need to be incorporated in the objective function and the weights need to be optimized for increasing number of product quality constraints.

Particle swarming appears to be a promising approach to optimize the yield of a complex coal plant. The future prospect of the study lies in modifying the objective function to one unique function that can incorporate any quality constraints with no reward / penalty on the cost components.

ACKNOWLEDGMENT

The authors express their sincere thanks to Illinois Clean Coal Institute (ICCI) and the Illinois Department of Commerce and Economic Opportunity for funding the project without which the work could not have been conducted. The authors also express their thanks to the American Coal Co. for their support to the project and also to the personnel of Galatia Coal Preparation Plant for helping us in collecting samples, and the full support for the completion of this study. Sincere acknowledgements are extended to Dr. S.K. Biswal, scientist, Regional Research Laboratory, Bhubaneshwar, India, and all colleagues for their support through the course of the project.

REFERENCES

Bergh, F.V.D. and Engelbrecht, A.P. (2004). A Cooperative Approach to Particle Swarm Optimization. *IEEE Transactions on Evolutionary Computing*, 8 (3), pp. 225-239.

Fisher, R.A. (1936). The use of Multiple Measurements in Taxonomic Problems. Annals of Eugenics, 7, 179-188.

Gupta, Vishal, Mohanty, M.K., Mahajan, Ajay and Biswal, S.K. (2004). Performance Optimization of a Coal Preparation Plant Using Genetic Algorithms," *ASME International Mechanical Engineering Congress*, Anaheim, California.

Gupta, Vishal, Mohanty, M.K., Mahajan, Ajay and Biswal, S.K. (2005). Genetic Algorithms – A Novel Technique for Optimization of Coal Preparation Plant," *International Seminar on Mineral Processing Technology*, Dhanbad, Jharkhand, India.

Honaker, R. Q., Sevim, H., Mohanty, M. K., Patwardhan, A. and Rajan, B. G. (1997). Advanced fine coal processing for improved mine profitability. *Final technical report, Illinois Clean Coal Institute*, Project Number 96-1/1.1A-1, Carterville, Illinois.

Kennedy, J. and Eberhart, R. (1995). Particle Swarm Optimization. *Proc. IEEE International Conference on Neural Networks*, IV, Washington DC, USA, pp. 1942-1948.

Luttrell, G.H., Honaker, R.Q. and Yoon, R.H. (2004). Optimization of the coal fuel supply chain: a coal preparation perspective. *29th International Technical Conference on Coal Utilization,* 2, Clearwater, Florida, USA, pp. 1356-1364.

Lyman, G.J. (1993). Computational procedures in optimization of beneficiation circuits based on incremental grade or ash content," Trans. Inst. Min. Metall., pp. C159-C162.

Mendes, R., Kennedy J. and Neves J. (2004). The Fully Informed Particle Swarm: Simpler, Maybe Better. *IEEE Transactions on Evolutionary Computation*, 8 (3), 204-210.

Pomeroy, P. (2003). An Introduction to Particle Swarm Optimization. http://www.adaptive view.com.

Salama, A.I.A, 1991, Graphical techniques for estimating the separation characteristics based on mass distribution. *Coal Preparation*, 9, pp. 51-63.

Salama, A.I.A. (1986). Yield maximization in coal blending. *10th Intl. Coal Preparation Congress*, Edmonton, Canada.

Salama, A.I.A. (1998). Coal washability characteristics index utilizing the M-Curve and the CM-Curve. *Intl, J. of Mineral Processing*, 55, pp. 139-152.

Sarkar, G.G., Sakha, S., Lahiri, A. (1960). A Graphical Method of Determining Optimum Operating Conditions for Washing Mixed Coals. *Journal of the Institute of Fuel*, 33, pp. 230-240.

In: Characterization and Development of Novel Materials
Editor: Rafiq Islam
ISBN: 978-1-61324-149-3

Chapter 3

Development of Probabilistic Simulation Model for Production Optimization in Open Pit Mines

M. Govinda Raj*[*,1], Harsha Vardhan[1] and Y. V. Rao[2]

[1]Department of Mining Engineering
National Institute of Technology Karnataka, Surathkal
P.O. Srinivasnagar – 75025 (D.K)
Karnataka, India
[2]National Institute of Technology Warangal
Warangal – 506004
Andhra Pradesh, India

Abstract

Large mechanized open pit operations can become more efficient and competitive without additional costs by improving the productivity of the existing machinery. This can be achieved by ensuring higher availability and utilization of the equipment The present paper deals with truck dispatch system using a variety of strategies to enhance the production capacity of a given set of equipment (shovels and trucks) by way of continuous dispatching of trucks. A discrete event probabilistic simulation program has been developed to demonstrate the production potential of an operating mine. It can accommodate any number of shovels and trucks of different capacities. The characteristic features of each truck and shovel can be individually specified to reflect the reality of actual working conditions. The model was tested on a hypothetical mine and a real case to verify the logic used in the program development. Subsequently, the model was implemented in two real mines to investigate the production potential. It was observed that substantial scope exists for productivity improvement when the mines were operating on a flexible mode.

---

* Email: govindaraj88@yahoo.co.in.

Keywords: Open pit mines; Production optimization; Probabilistic simulation model; Shovel-truck system; Productivity improvement of machinery.

1. INTRODUCTION

The development of a simulation model for any production function aids in a better understanding of the system and provides useful insights into its functioning. The simulation model would be of benefit, at the design stage, in actual operation planning and management and for future expansion of an existing operation. A number of events take place in a mine at different time intervals that are not amenable to strict mathematical modeling and analysis. Under such situations the simulation technique provides a means for a systematic study and investigation of the phenomena for useful insights regarding its working. It also reveals the bottlenecks that may restrict the achievement of full potential in terms of production and other system parameters.

A large open pit shovel-truck system can be appropriately characterized as a "highly complex man-machine system". The following features evidence its complex nature:

- The changing topography of the pit and the associated network of haul routes.
- The status of the operating equipment like idle, busy, breakdown, maintenance etc.
- The limited capacities of crushers and certain waste dumps.
- The three dimensional variation of ore quality at the shovels.
- Pit planning, technical, and economic considerations that govern the quantum of ore and waste removal per shift.
- Management attitudes, goals and pre-dispositions regarding the achievement of certain goals.

Under the above mentioned set of circumstances, decisions regarding the effective utilization of shovel-truck resources are of critical importance for achieving maximum production and any other relevant managerial goals like quality, ore/waste ratios etc.

The present research investigation was undertaken to find better/improved ways of mine operations management, with the following objectives:

- Development of a fully functional probabilistic discrete event simulation model of an open-pit mine.
- Testing and validation of the developed model.
- Model implementation through case study.

Analysis of the developmental history of mine system simulation reveals that since its beginning in the 1950's to its current level of evolution, there has been a gradual refinement of objectives, approaches and focus areas of attention. Thus, in some sense a number of researchers have made incremental improvements in various dimensions of simulation modeling. The present investigations are one such incremental addition efforts.

Many significant decisions taken by the mine management are generally based on technical suitability and economic analysis and predetermined rating systems and tables. As financial and other constraints on mining operations become more severe, more accurate methods are required like discrete event simulation programs and other approaches. A variety of discrete event simulation programs have been used successfully in many mining applications. Achieving optimality for a number of dynamically varying factors which are critical for mine functioning has been a goal which has been addressed to varying degrees of success by a number of researchers (Chironis, 1985; White and Olson, 1986; Bonates and Lizotte, 1988; Sturgul, 1999; Dessureault et al., 2001; Michalakopoulos, 2001; Nelson and Peterson, 2002; Alarie and Gamache, 2002; Awuah-Offei et al., 2003; Ramazan and Dimitrakopoulos, 2004; Cohen et al., 2004; Lewis et al., 2004; Huang and Espley, 2005; Burt and Caccetta, 2007; Askari-Nasab et al., 2007; Gholamnejad and Osanloo, 2007).

2. Methodology of Development of Probabilistic Simulation Program

The development of a simulation model for any real system needs to consider the nature of the variables that describe the behavior of the system. A certain change in the values of the system variables leads to a certain change in the behavior of the system termed as "system response". The variables that comprise the elements of the system may be characterized by either discrete or continuous distribution.

A probabilistic system is one that comprises more than one random variables that change simultaneously. This calls for describing the behavior of each variable in terms of a certain distribution having a certain probability of occurrence. The main events that occur in such a system are the start and end of loading a truck, the start and end of travel (full), the start and end of unloading, the start and end of waiting at the crusher, the start and end of travel (empty), the start and end of waiting at the shovel. The durations of these events give the loading time, travel (full) time, travel (empty) time, waiting (at a crusher) time, unloading time, travel (empty) time, and waiting (at shovel) time. As these times are subject to random variations, the representation of these events should be in terms of probability distributions. The software system developed has been designed to perform either a deterministic simulation or a probabilistic simulation. The deterministic simulation is only indicative of the probable behavior of the system where as the probabilistic simulation almost replicates the behavior of the real world system. The developed simulation program incorporates shovel operation modeling, truck operation modeling, crusher/dump modeling, breakdown events modeling and cost computation modeling.

2.1. General Structure of the Simulation Program

The simulation program is structured, on an “advance clock approach”, wherein the clock advances not by a “fixed-time increment”, but by the “minimum event time”. This minimum event time is iteratively reduced, in order to generate the next minimum event. The development of the simulation algorithm in this way, leads to the capturing of every event that occurs in the system, however small the time duration may be. This way of designing the algorithm ensures that no data/information is lost, thus ensuring the highest level of accuracy for tracking the "time of occurrence" of every event. The general structure of the simulation program is shown in Figure 1.

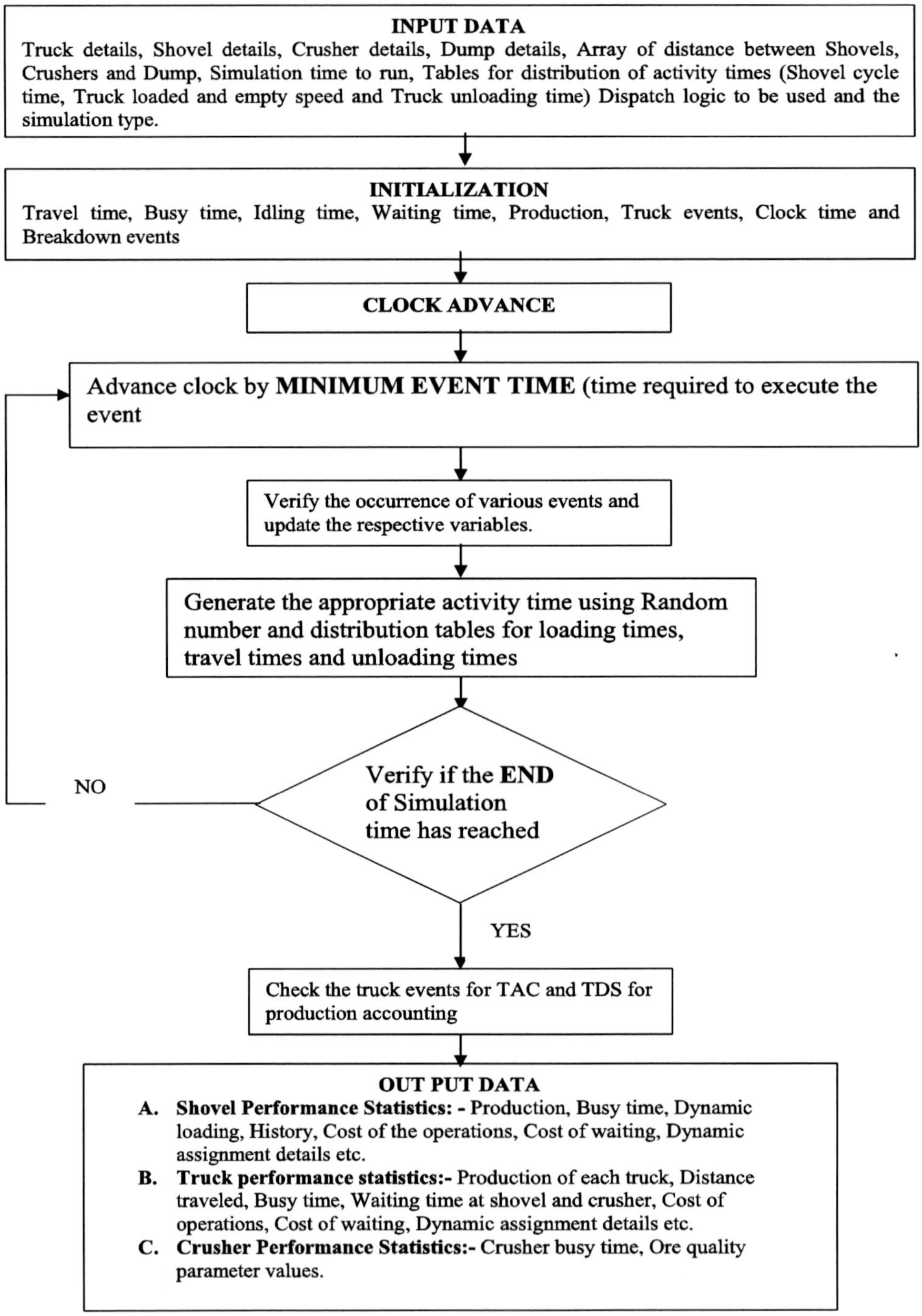

Figure 1. General structure of the simulation program.

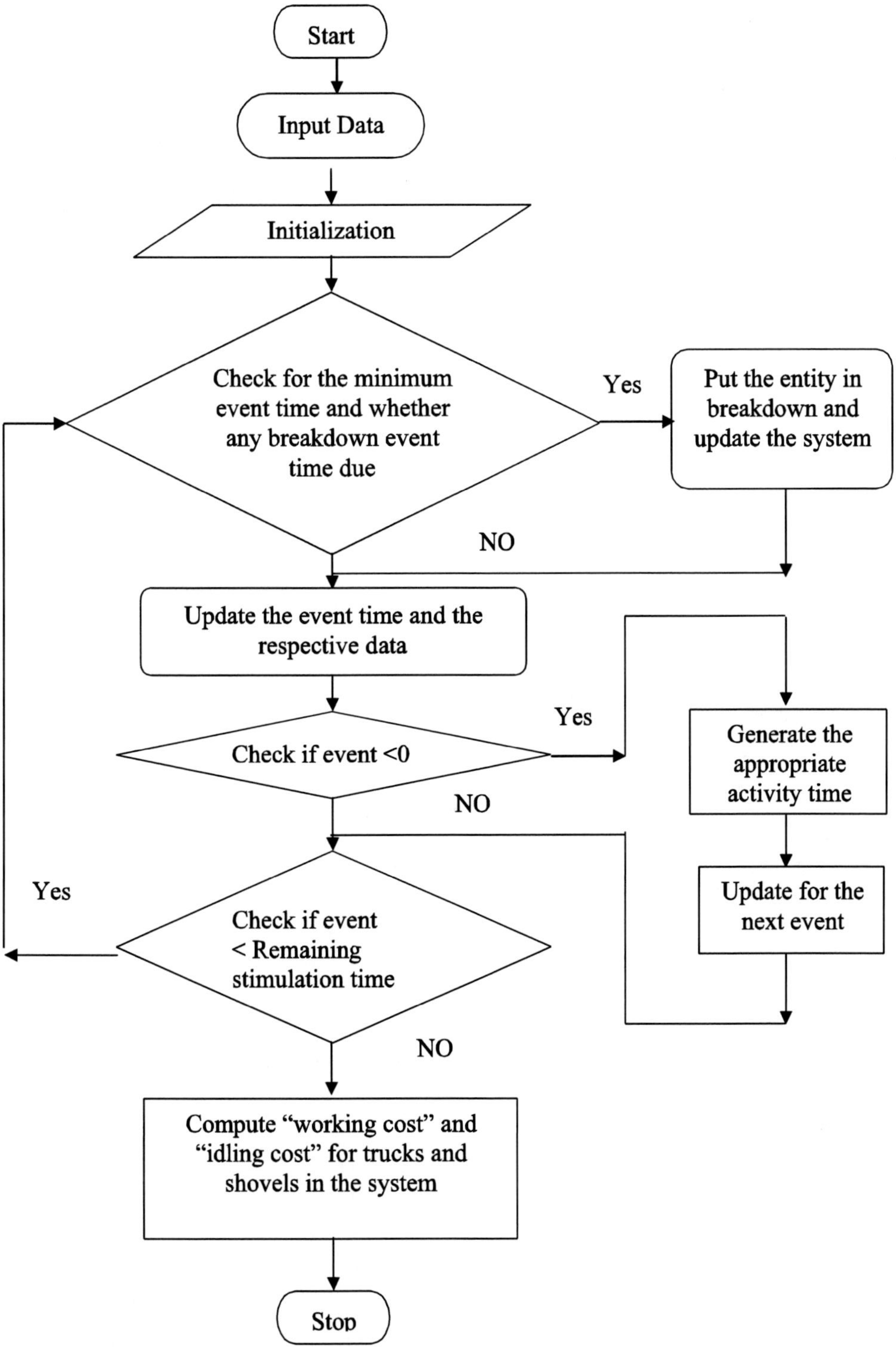

Figure 2. Simplified flow diagram of the event oriented simulation program.

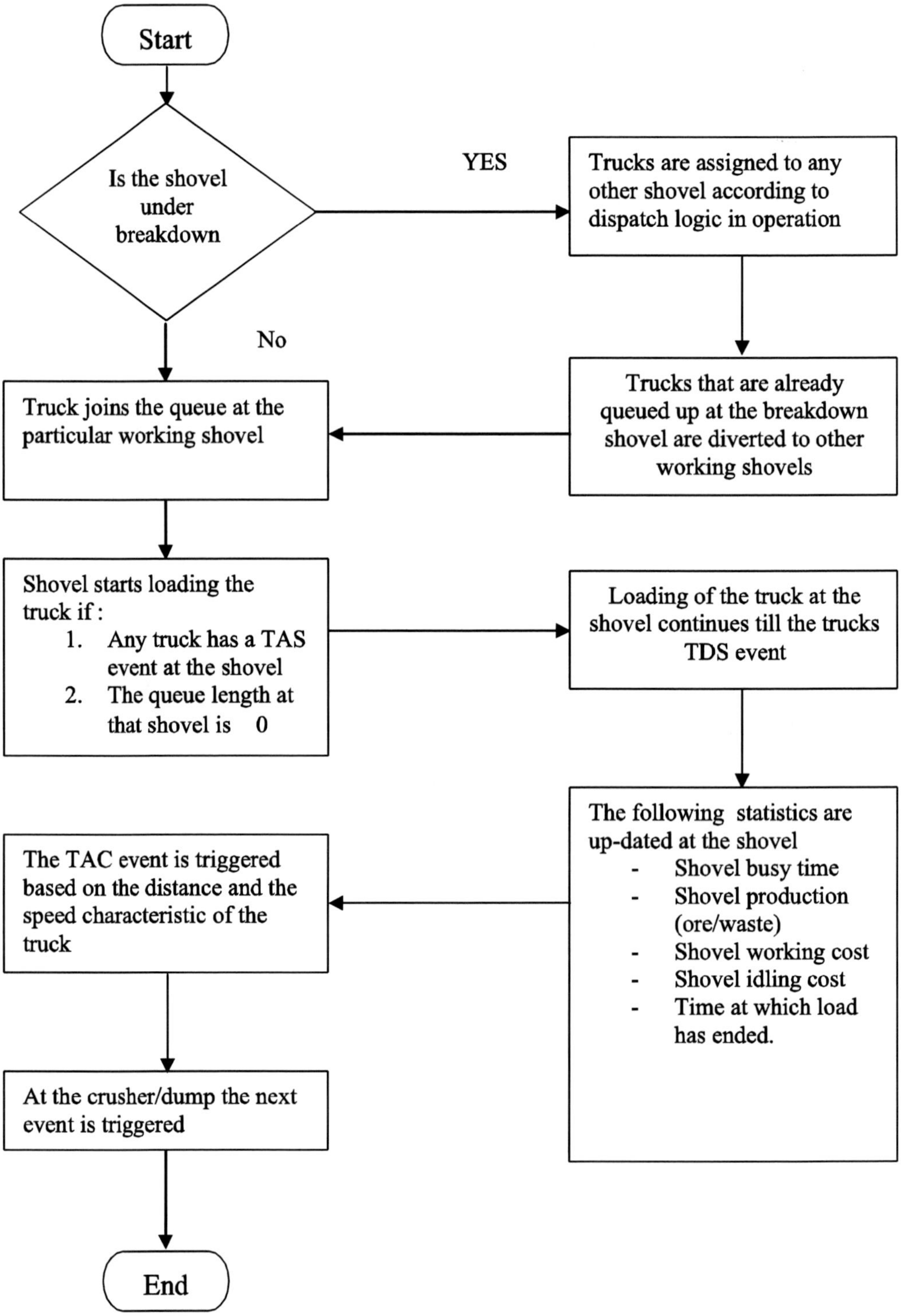

Figure 3. Flow diagram of processes at a shovel.

The design of the simulation program can be based on either "time oriented modeling" or "event oriented modeling". In order to ensure that every event occurring in the system is tracked and captured, the event oriented approach was selected with a slight modification. The simulation time advances to the next imminent minimum event. At this state, all the relevant statistics of the system are updated. In other words, data updating is done on an event by event basis. The simulation program can be interrupted, at any stage to have an idea of the system statistics, at that instant. Figure 2 depicts the simplified flow diagram of the event oriented simulation program.

2.1.1. Shovel Operation Modeling

The modeling of the shovel operation is based on the individual characteristics of the particular shovel. Figure 3 shows all the processes that take place at a shovel. Before a truck is dispatched to a particular shovel, the system checks for its operational status.

If it is in active state the TAS (Truck Arrival at Shovel) and TDS (Truck Departure from Shovel) events are tracked to capture the loading time for that truck. This process also computes the waiting time of the truck at that shovel. The “truck queues” at the various shovels can be tracked and waiting time of the trucks can be updated regularly. Thus at each shovel, after a truck has been loaded the TDS event is tracked and the following variables are updated: tonnage loaded by that shovel, the busy time of the shovel, the working cost of the shovel, number of trucks loaded by that shovel. The TDS event triggers the next TAC (Truck Arrival at the Crusher/Dump) event, based on the distance between the shovel and the crusher/dump.

2.1.2. Truck Operation Modeling

The trucks in the mine can be of different makes, types and capacities. The system allows the addition or deletion of the trucks, each having a unique identification number and characteristics. The spotting time at a shovel and crusher, the unloading time, the travel speed both loaded and empty need to be specified.

The trucks are initially allotted to specific shovels based on the target production of each shovel and the lead distance to the unloading point. The movement of the truck entity is modeled as a process having one of the four states, namely Truck Arrival at Shovel (TAS), Truck Departure from Shovel (TDS), Truck Arrival at Crusher (TAC) and Truck Departure from Crusher (TDC).

The TAS event specifies the time at which the truck would arrive at a particular shovel. This gives the duration of travel (empty), based on the path of travel, truck speed and the dispatch logic in use. The TAC gives the time at which the truck will arrive at the crusher/dump, thus giving the duration of travel (full) depending on the truck capacity, truck speed and the distance between the shovel and crusher. The TDC event specifies the time at which the truck would complete its unloading into a crusher/dump. This includes the waiting time at the crusher and the actual unloading time of the truck with the spotting time at crusher. In the case of a dump, TDC denotes only the spotting and unloading time based on truck identification number, as there is no waiting time associated with unloading at dump points. The TDS gives the time at which the truck would be loaded at a particular shovel and ready for the next event. TDS and TAS taken together for a shovel and truck provide the loading time of the shovel and truck waiting at the shovel, if any. From the four states, the loading time of a truck, the travel (full) time, travel (empty) time and the unloading activities can be computed. Each truck is assigned a particular shovel, once it completes unloading and is ready for dispatch, according to the logic being used.

2.1.3. Crusher/Dump Modeling

The model can incorporate any number of crusher/dump points. In the case of a crusher, all loaded trucks originating from ore shovels are tracked, and their TAC events are arranged in an ascending order. The crusher is unavailable for any other truck till a truck currently spotting/unloading is completed. The "first in first out" (FIFO) queue discipline is followed and the waiting time of the trucks in queue at the crusher is updated after every event. In the case of a dump, the trucks are allowed to spot and unload simultaneously, as is generally the practice in most open pit mines. According to the dispatch logic in use, the trucks are assigned to a shovel, on completion of the unloading activity.

2.1.4. Breakdown Events Modeling

The breakdown of a truck in the model makes the truck move out of the simulation for the breakdown duration. The truck under breakdown is recorded to take care of production accounting as well as updating the relevant statistics. A truck in loaded state is considered for production accounting and material balancing purposes, in case it is still in breakdown state, till the end of simulation period.

In the case of a shovel breakdown, the number of processes involved is quite large. At the instance when the shovel breakdown is due, the model prompts for re-allotment of trucks that are allotted to it in the case of fixed circuit operation. In the case of flexible circuit under any dispatch logic, the system models the following processes:

- The shovel number under breakdown is recorded, and the trucks are not dispatched to this shovel, till it comes back to an active state.
- The trucks, which are in queue at the shovel, are transported back to the crusher/dump for reassignment to other shovels.
- The trucks traveling empty towards the shovel are brought back to the crusher/dump for reassignment.

2.1.5. Cost Computation Modeling

The total number of shovels and trucks present in the simulation model are given a cost attribute comprising ownership cost, operating cost (working) and operating cost (idling). These costs are computed at the end of the simulation run from the busy times and idling times of each individual equipment. This enables the actual assessment of the equipment performance in a mine in monetary terms of those equipments which actually participates in the production activity and contributes to the final production. The shovels and trucks in the system that may be in an idling/waiting state also incur an operating cost (waiting) and hence do not contribute to production for the duration of idling.

The production given by a piece of equipment, and the associated cost will facilitate a comparative normative evaluation of the system under different operational scenarios.

2.2. Details of Software System Development

The discrete event probabilistic simulation program has been designed and developed, keeping in view the needs of a wide variety of open pit mines. The database for the system comprises the list of shovels and trucks being used, each identified by a unique identification number. As mines generally operate with two or three different varieties of shovels, having different bucket capacities, the system enables the incorporation of the specific features of each machine separately.

The dispatch logics used can be either modified or added as a separate module, so that the program offers the required flexibility for future expansion and enhancement. The simulation model is based on incrementing the simulation time by "minimum event time", so that each and every event occurring in the system is captured and all the relevant processes are performed on the occurrence of that event. Thus, there is no loss of accuracy, as it generally happens, when an advance clock approach with a fixed time increment is taken.

The general purpose *VISUAL BASIC* language has been used in the development as it offers sufficient flexibility to model any event. In addition, the excellent debugging facilities offer scope for faster model development and testing. The software system developed has a database to include any number of shovels, trucks, crushers, waste dumps and specific ore quality parameters at the places in the mine, where ore shovels operate. This enables the software to be a general purpose one that can be tuned to the needs of any mine. The dispatch logics used in the system can also be supplemented as the program has been developed on a modular basis.

The software system comprises graphical animation module that is driven by the current simulation data. Thus a real-time animation of the simulation process can be viewed, which enhances the understanding of the working of the system. Whenever a truck or shovel breakdown event is due, a color code has been given to indicate the machine status.

The animation module allows specification of the location of shovels, dumps and crushers based on a co-ordinate system. The actual field distance and the pictorial distance are not mathematically correlated, in order to make the model, more easy to use. A "control array" is used to dynamically generate the controls for trucks, shovels, crushers and dumps. The movement of trucks from one position to another is controlled by the line equation governing the truck and shovel (origin position) and the truck and dump/crusher (destination position). The crusher location is taken as the base location (origin) and the shovels as destination for dynamic dispatch situations.

As every truck can have four states namely TAS, TAC, TDS, and TDC, the program checks for the next occurring event of each truck and performs the necessary computation to either move or stop. The individual truck numbers are displayed on the truck icons for easier identification and tracking of the truck movement. Whenever a truck/shovel breakdown is over, the truck will start moving towards their respective destinations, according to the dispatch logic being used.

The simulation model requires clarification regarding what the 'events' are and 'how' they are used in the model. The program tracks the occurrence of the four main events namely TAC, TDC, TAS, and TDS. These four events occurring any where in the mine are continuously monitored to enable updating of the respective variables and data. A precise clarity regarding the duration of each event enables the standardization of the data collection program.

Representative data were collected for each type of loading shovels regarding its loading time and for each type (capacity) of truck, about the spotting time (at shovel/crusher/dump), unloading time (at crusher/dump) and the average travel speeds. As these data are stochastic in nature the model provides two types of input format:

- Data given as a time distribution curve for each activity.
- Data fed into a database as it is collected on a specific day.

The second approach enables a more realistic portrayal of the field conditions, as it existed on a specific working day. In addition, this has the advantage, that field data which do not fit into any standard mathematical distribution also can be very easily modeled. Production modeling through this approach, it is believed, would enhance the portrayal of "real field conditions", as it existed on the day and time of data collection.

3. Testing and Validation of the Model

Testing and validation is an essential component of the model development. The model was tested on a hypothetical mine and a real case to verify the logic used in the program development.

3.1. Hypothetical Mine

The probabilistic simulation model and the software system, were first tested and verified for a hypothetical mine in order to check the accuracy of the system logic that has been used and incorporated into the model.

A hypothetical mine working with four 10 m^3 shovels and eight 85T dump trucks were chosen. The haul road network, the loading capacities of the individual shovels, the operational characteristics of the trucks (speed, spotting times etc.) were considered for computation of system productivity and calculation of the loading and hauling expenditure. The operation of the mine was tested for both locked-in system and for a variety of dispatch criteria. In order to evaluate the increase in the system productivity for an increase in the number of trucks in operation, simulation experiments were performed utilizing 8, 10, 12, 14, 16 and 18 numbers of trucks in operation.

The results revealed that the system reaches a saturation level between 14 and 16 number of trucks operating in the mine. A further increase in the number of trucks increases the waiting time without any benefit in production improvement. Further, the shovel utilization level remains constant thereafter. An examination of the busy time of the trucks and the waiting time (at the shovel/crusher) provide further evidence of system saturation. Table 1 gives the results of simulating the system starting from 8 to 18 trucks for the fixed circuit operations.

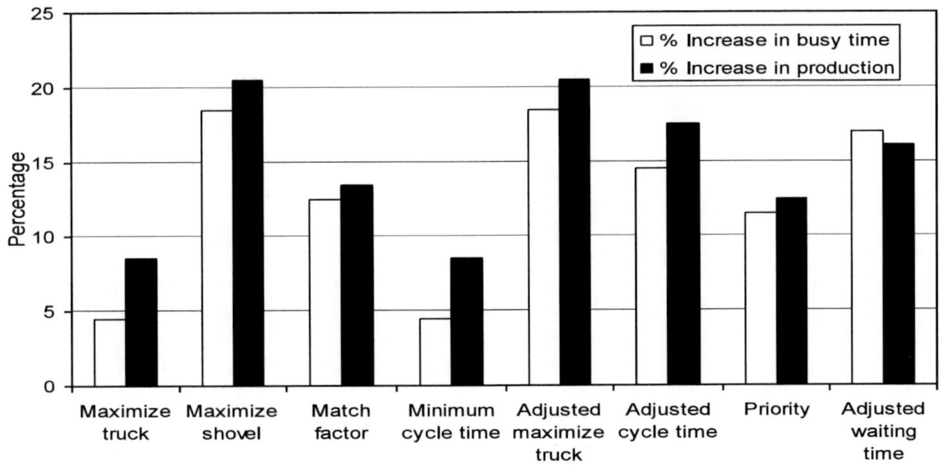

Figure 4. Comparison between percentage increase in production and truck busy time for 12 trucks in the system (Hypothetical mine model).

The operation of the mine under flexible dispatching system revealed that there is considerable scope for productivity improvement. For instance, using the maximizing truck use logic, the production obtained was 22,002 tonnes which was an increase by 10.57 %. Table 2 gives the improvement in system production for some of the dispatch logics tested. Figure 4 compares the percentage increase in production and the busy time of the trucks for various dispatch logics.

Table 1. System productivity for a hypothetical mine (Locked-in Operation)

No. of trucks	8	10	12	14	16	18
System productivity (tonnes)	18,089	19,915	22,661	25,024	28,363	28,363
Shovel number 1(tonnes)	3,870	4,611	4,611	4,611	4,611	4,611
Shovel number 2(tonnes)	4,004	5,089	5,089	5,089	5,089 I	5,089
Shovel number 3(tonnes)	5,490	5,490	8,236	8,236	9,447	9,447
Shovel number 4(tonnes)	4,725	4,725	4,725	7,088	9,214	9,214
Trucks busy time (hours: minutes: second)	47:37:26	53:09:40	59:06:32	65:06:58	73:10:43	73:10:42
Waiting time at crusher and shovel (hours: minutes: second)	0:32:21	7:04:06	13:10:57	19:17:46	23:15:47	35:16:07

Table 2. Increase in percentage production for different dispatch logics over fixed logic (Hypothetical mine)

No. of trucks	Maximize truck	Maximize shovel	Match factor	Minimize cycle time	Adjusted Maximize truck	Adjusted cycle time	Priority	Adjusted Waiting time
8	3.23	0.80	1.26	3.23	5.96	4.62	2.00	3.81
10	10.47	16.52	12.72	4.18	16.47	14.19	10.01	13.39
12	5.58	20.37	13.39	8.58	20.38	17.33	12.66	16.26
14	10.69	13.34	13.01	9.37	13.34	12.02	12.37	12.07

Table 3. Actual Mine Production and Simulated Mine Production

Shovel number	Actual mine production (Tons)	Simulated production by model (Tons)	% Error
1	2,000	2,024	+1.500
2	624	659	+5.600
3	932	880	-5.570
4	6,159	6,182	0.003
5	6,495	6,443	0.008
6	4,430	4,443	0.002
7	2,799	2,630	6.030
TOTAL	23,439	23,261	0.007

For the hypothetical mine network, the maximize shovel and the adjusted maximize truck logic gives the highest improvement. It was also observed that production improvement depends on the dispatching methodologies which may vary from mine to mine as each mine is unique in terms of layout, machinery used and the management's operational philosophy. The dispatch logic can be modified and tailored to visualize and realize the maximum benefit, for a given set of field conditions.

3.2. Real Mine

Table 3 gives the production statistics for a shift in a real mine for the month of June 2001. The actual production obtained by each individual shovels, the working duration of the shovel, and the number of trucks allotted/used in the mine was considered. It was decided to simulate the model for a duration of 6 hours, representing the working of one shift. The actual production obtained from the mine and the average production given by the simulation model, are given in Table 3. The results of mine production obtained through the simulation model are almost inline with that of actual mine production.

4. Case Study and Model Implementation

The probabilistic simulation model is primarily intended to analyze an existing mining operation. As the mines are operating on a fixed /locked in mode, there is a large amount of under utilization of the shovel-truck resources. The idling time of the shovels and the waiting time of the trucks represent the quantum of potential lost production. The simulation model is first used to study the mine as it functions, by replicating the mine behavior through the simulation model. The simulation model is used with the same level of shovel-truck resources to investigate, alternative methods of operation. A variety of dispatch logics is used to understand the system dynamics, which would result in an increase/decrease of some parameters. This information provides the mine management useful insights regarding the quantum of benefit envisaged, by implementation of alternative modes of operation.

4.1. General Description of Mine -A

Mine-A is an open pit coal mine using shovel truck combination for extraction of coal and overburden removal. Loading and hauling operations of coal are done by two shovels of 5.0 m^3 capacity. The overburden removal is by means of three shovels, including two of 10.0 m^3 capacity and one of 5.0 m^3 capacity. The time study data were collected to know the details regarding average loading time for shovels of different capacities, the average travel speed of the trucks, the average travel time from a loading point to the unloading point etc. This information is used as the input data for simulating the mine. The mine had sixteen trucks in operation.

4.1.1. Current Production Scenario

The behavior of this mine was simulated for a working shift of six hours duration. The remaining two hours of time is due to various activities/events like lunch and tiffin breaks, maintenance work, shift start and end delays etc. The production simulation was attempted for the operation of a total of five shovels with sixteen trucks. The total production under fixed/locked in system of operation was 13,552 Tons.

The production model of the mine and the equipment cost model of the mine give the complete picture of the average utilization of each shovel and truck in the operation, as well as the cost of the operations. The total cost of truck operation and shovel operation gives the total cost of material handling for the mine. The truck and shovel cost models assume a set of economic conditions, under which they are developed and used.

4.1.2. Production Potential under Dispatching

The simulation model was used to investigate the behavior of the mines under a variety of dispatch strategies. It was observed that the system productivity could be increased considerably, with the same fleet of fixed equipment resources. The dispatch methodologies try to convert the idling time present in the fixed mode of operation into the useful production time. Thus the combined utilization of the shovel-truck resources can be considerably increased. This can be examined on an individual equipment basis, as well as on a system level.

Figure 5 depicts the scenario of production increase under different logics. For this mine, the simulation results show that the minimum cycle time dispatch logic yields the maximum increase in production when compared with the fixed mode of operation. There is an increase of 18.32 % in terms of production, and the smallest increase is for priority number dispatching of 7.49 %.

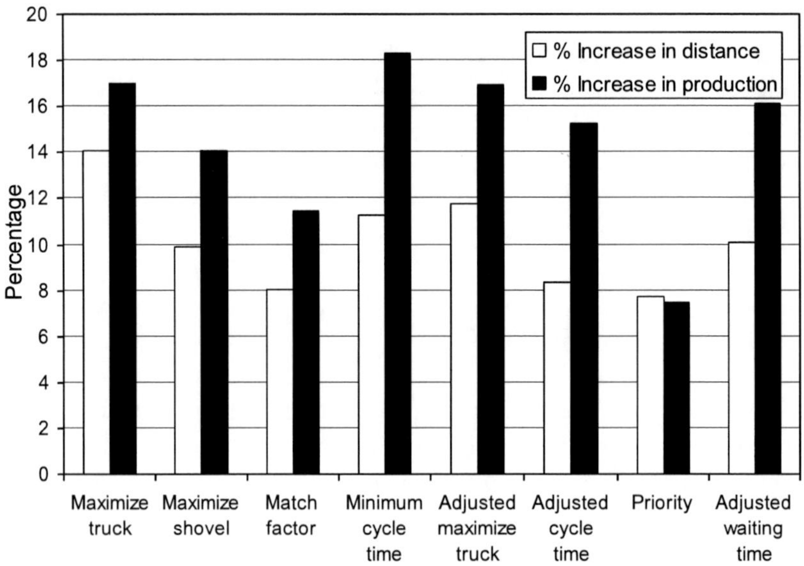

Figure 5. Percentage increase in production and distance traveled under different dispatch logics for Mine – A.

4.1.3. Analysis of Mine-A Performance

When the mine operates on a flexible/dynamic mode, the degree of benefit obtained depends on a number of system parameters like the capacity of the shovels and their location in the pit configuration, the individual truck cycle time between a shovel and the corresponding unloading point, the loading time of the shovel, the truck travel speeds, the spotting and unloading time of the trucks, the queue length at the crusher at various times during the shift etc.

Table 5. Percentage increase in production for different dispatch logics for Mine-A

S. No.	Dispatch logics	Production (Tons)	% Increase in production
1.	Fixed	13551	0.00
2.	Maximize truck	15854	16.99
3.	Maximize shovel	15457	14.07
4.	Match factor	15108	11.49
5.	Minimum cycle time	16034	18.32
6.	Adjusted maximize truck	15846	16.94
7.	Adjusted cycle time	15616	15.24
8.	Priority	14566	7.49
9.	Adjusted waiting time	15740	16.15

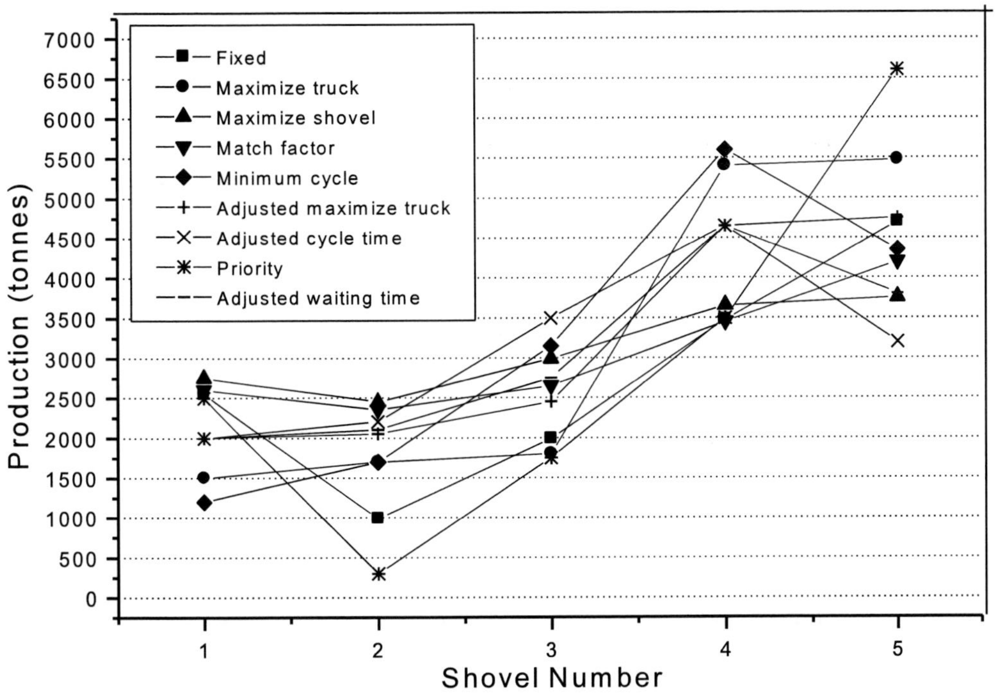

Figure 6. Variation of production under different dispatch logics for Mine – A.

An examination of Figure 6 shows that the production from shovel No. 2 increases considerably, once the system is made flexible. The maximize shovel dispatch logic gives a high production for this shovel. The shovels Nos. 4 and 5, working in the overburden, give the highest production for shovels under minimize cycle time dispatch logic and likewise under priority dispatch logic. The production variation under fixed and flexible system of operation is further explained by Table 4, wherein shovel No.2 has the highest idling time under the fixed mode of operation. This idle time is considerably reduced for maximize truck and match factor logic. The priority number dispatch logic has less production, due to the high idling times for shovel Nos. 2, 3 and 4.

Table 5 gives the percentage increase in production for different dispatch logics. The largest increase in system production is accounted by the shovels working in overburden, i.e., shovel Nos. 3 and 4 which show very high busy time in adjusted cycle time and minimum cycle time (Table 4). The very high production for the shovel No.4 is also accounted for in maximize truck logic.

From the perspective of the cost of loading and hauling operation, Table 6 shows that maximize shovel and adjusted cycle time decreased idling cost by 18 % as compared with the fixed mode of operation. The smallest reduction in the idling cost is obtained in the adjusted waiting time logic, with a corresponding increase of 16.15 % in the production over the fixed logic.

As regards the operating cost of the shovels, the priority number dispatching logic provides the lowest operating cost, which corroborates low production – increase (Table 7).

Table 6. Percentage decrease in idling cost and percentage increase in production under different dispatch logics for Mine – A

Dispatch logics	Percentage decrease in idling cost	Percentage increase in production
Maximize truck	11.60	17.00
Maximize shovel	18.01	14.07
Match factor	14.25	11.49
Minimum cycle time	15.60	18.32
Adjusted maximize truck	15.40	16.94
Adjusted cycle time	18.06	15.24
Priority	1.39	7.49
Adjusted waiting time	1.17	16.15

Table 7. Shovel operating costs (in thousands of rupees) under different dispatch logics for Mine – A

Dispatch logics	Shovel Number				
	1	2	3	4	5
Fixed	10.5	4.1	7.7	10.3	11.3
Maximize truck	6.1	7.7	7.0	16.2	13.1
Maximize shovel	11.1	11.0	11.8	10.8	9.0
Match factor	10.3	10.2	10.2	10.2	10.1
Minimum cycle time	5.1	7.6	12.2	16.8	10.3
Adjusted maximize truck	8.1	9.0	9.7	14.1	11.4
Adjusted cycle time	8.1	9.9	13.8	14.1	8.1
Priority	10.0	1.6	7.0	10.6	15.9

Table 8. Sample distribution of shovel costs (in thousands of rupees) for different dispatch logics for Mine-A

Shovel No.	Fixed			Maximize truck			Maximize shovel		
	Operating cost	Idling cost	Total cost	Operating cost	Idling cost	Total cost	Operating cost	Idling cost	Total cost
1	10.40	6.72	36.12	6.17	10.65	35.81	11.32	5.87	36.18
2	4.20	12.48	35.67	7.68	9.26	35.92	10.86	6.30	36.15
3	7.71	9.22	35.92	6.99	9.89	35.87	11.69	5.52	36.21
4	10.42	11.24	45.47	16.50	5.71	46.01	10.97	10.75	45.52
5	11.41	10.34	45.55	13.26	8.66	45.72	8.97	12.56	45.34
Total	44.15	50.01	198.74	50.59	44.17	199.34	53.81	41.00	199.39

Table 9. Distribution of truck distance traveled for different dispatch logics for Mine–A

	Distance traveled, km		
Dispatch logics	Full	Empty	Total
Fixed	387	377.36	764.36
Maximize truck	440.24	432.60	872.84
Maximize shovel	424.82	415.10	839.92
Match factor	418.86	406.82	825.68
Minimum cycle time	433.64	417.80	851.44
Adjusted maximize truck	435.66	419.22	854.88
Adjusted cycle time	419.28	408.68	827.96
Priority	416.92	406.38	823.30
Adjusted waiting time	425.52	416.67	842.19

The minimum cycle time shows the highest operational costs, which are due to the more number of actual hours of operation that translates into an increase in production. Table 8 gives the distribution of shovel costs for fixed, maximize truck and maximize shovel logics.

Table 10. Variation of truck costs for different dispatch logics for Mine – A

Dispatch logics	Cost in thousands of Rupees		
	Operating cost	Idling cost	Total cost
Fixed	170.0	10.0	320.0
Maximize truck	200.0	9.8	345.0
Maximize shovel	200.0	9.6	345.0
Match factor	195.0	10.1	345.0
Minimum cycle time	200.0	9.9	349.0
Adjusted maximize truck	200.0	9.7	347.0
Adjusted cycle time	200.0	9.7	346.0
Priority	180.0	11.0	345.0
Adjusted waiting time	200.0	9.8	345.0

The performance of the trucks can be evaluated in terms of the total distance traveled, more specifically the loaded distance traveled, as it directly adds to the production. Table 9 gives the actual distances traveled under different logics. In comparison with the fixed logic, the cumulative distance traveled by all the trucks is much higher, which is reflected in the production increase in each of the dispatch logics. Because of the higher utilization of the trucks for production activity, the operating cost of the trucks is considerably higher where as the idling time and idling cost of the trucks varied marginally (Table 10).

The trucks in the system either wait at the shovel or at the crusher and this is considerably reduced for maximize shovel, adjusted maximize truck, adjusted cycle time and adjusted waiting time. The match factor dispatching method shows higher values both at the crusher and the shovel.

4.2. General Description of Mine - B

Mine - B is a large open pit coal mine utilizing shovel dumper combination. Coal loading was done by three 5.0 m^3 shovels. Overburden was removed by two shovels of 10.0 m^3, one of 5.0 m^3 and another of 3.3 m^3 bucket capacity. Twenty two trucks were in operation in the mine under the fixed mode of operation. The trucks were of 35, 50 and 85 Tons capacity.

4.2.1. Current Production Scenario

The mine was operated in the fixed mode, and hence the time study data collected was used to first replicate the simulated mine production to the actual production. The simulation was done for duration of six hours, out of actual shift duration of eight hours. The mine production model and the equipment cost model were used to simulate the actual production operations. The combined system output for coal and overburden was 23,439 Tons for a representative one shift of mine working.

4.2.2. Production Potential under Dispatching

The application of the simulation model to examine the production potential under a variety of dispatching methodologies revealed some interesting relationships among different system parameters. As this was a relatively large mine, wherein the equipment were dispersed over a wide area, the degree of production benefit is considerably lower as compared with Mine – A. Nevertheless, dispatching provides an increase in the equipment utilization levels and less waiting time for the trucks in the system. The combined utilization of the shovel-truck resources can be examined on an individual equipment basis, as well as on a system level.

The total system production, shovel busy and idling time and haulage costs for different dispatch logics for Mine – B are given in Table 11. As regards the individual shovel production levels is concerned, the minimum cycle time dispatching methodology has shown a significant increase for shovel Nos. 1, 2, 3 and 5 whereas it reduced for other shovels. This is due to the considerably long cycle time, for those individual shovel circuits. The maximum truck dispatching logic has shown a considerable increase in the case of shovel Nos. 3, 4 and 5 in comparison with fixed circuit.

The adjusted cycle time logic shows the highest increase in shovel busy time, followed by maximum shovel and adjusted waiting time logic. For this mine, it was observed that the maximum truck and priority number dispatching logics did not show much increase. Further, the above mentioned dispatching methods show the least amount of shovel idling time.

From the perspective of cost analysis of the various dispatching methodologies, it was found that the truck costs are very high for the match factor and priority number dispatching methodologies. The truck haulage costs were found to be the lowest for minimum cycle time, and adjusted waiting time methods of dispatching. As regards the shovel total costs, it was found that the maximum truck dispatching logic accounts for the highest expenditure. When the production increase is also considered, this translates into a lesser per unit cost of operation.

Table 11. Total system production, shovel busy and idling time and haulage costs for different dispatch logics for Mine – B

Dispatch logics	Total system production (Hundreds of tons)	Total shovel busy time (Hours)	Total shovel idling time (Hours)	Total haulage costs (Thousands of Rupees)
Fixed circuit	232	18.4	23.7	527.0
Maximize truck	239	18.0	23.9	527.2
Maximize shovel	238	23.5	18.4	526.8
Match factor	233	21.9	20.2	528.6
Minimum cycle time	237	22.2	20.0	524.2
Adjusted cycle time	240	24.0	17.8	524.9
Adjusted waiting cycle	240	22.4	18.4	524.3
Adjusted maximize truck	244	20.2	21.6	528.7
Priority	217	17.8	27.2	528.2

4.2.3. Analysis of Mine -B Performance

In this mine being operated on a flexible dispatch mode, the dispatch logics that have been used have revealed that the general topography and current equipment location do not provide a very high benefit. The general conception that a dispatch system would increase the system production considerably may not be true for all mines. As the shovel locations change, the network of haul roads distances increase and the cycle time of individual circuits change. The unloading points for ore and overburden may also change and then facilitate a higher degree of benefit. An interesting observation here is that, the system benefit that results because of dynamic dispatching can be varied, by suitably modifying some of the key parameters like circuit cycle times and thereby the unloading points for ore and overburden.

CONCLUSION

The focus of the present work was to develop a fully functional probabilistic simulation model for analyzing open pit mines. The model can be used to structure and model a mining operation, so that a variety of simulation experiments can be conducted to analyze the system behavior and thus aid in decision making. The model has been designed to be general purpose in nature in order to have wide applicability. The primary aim of this research was to develop a mine production model and an equipment cost model and integrate these two, which can be used as a tool to analyze an open pit mining operation. The program allows for modifications to be incorporated in order to make it as a general simulation tool. A variety of heuristic dispatch methodologies have been incorporated for analyzing the productivity potential. The current model has provision to simulate equipment breakdown scenarios. Simulation driven animation program supplements the visual appeal.

The probabilistic simulation model developed is very useful to analyze the production scenario in a mine. Simulation experiments conducted with a model on mine data provided information on how the mine operations can be improved. The productivity potential of the mines was analyzed to develop new ways of operating a mine. When the model was implemented, it was observed that the system productivity could be increased considerably, with the same fleet of fixed equipment resources. The dispatch methodologies try to convert the idling time present in the fixed mode of operation into the useful production time. Thus the combined utilization of shovel truck resources can be considerably increased. This can be examined on an individual equipment basis, as well as on a system level. For instance, there was an increase of 18.32 % in terms of production under the minimum cycle time dispatch logic and the smallest increase was for priority number logic of 7.49 %, for Mine-A. In respect of Mine-B, the production increase was lower than Mine-A, indicating that the configuration of the mine haul road network and other parameters influence the "overall system benefit" that is possible. It is thus possible to investigate a number of "what if" scenarios/questions regarding many aspects of mine working, thereby serving as a handy tool to mine planning and design personnel.

ACKNOWLEDGMENT

The authors sincerely thank the reviewers who went through the paper in depth and provided valuable comments. They also thank the management of the mine for their kind cooperation in extending necessary facilities for carrying out this work.

REFERENCES

Alarie, S. and Gamache, M. (2002); "Overview of solution strategies used in truck dispatching systems for open pit mines"; *International Journal of Mining, Reclamation and Environment;* Vol. 16(1); pp. 59-76.

Askari-Nasab, H. and Frimpong, S. (2007); "Modelling open pit dynamics using discrete simulation"; *International Journal of Mining, Reclamation and Environment*; Vol. 21(1); pp. 35-49.

Awuah-offei, K., Temeng, V. A. and Al-Hassan, S. (2003); "Predicting equipment requirements using SIMAN simulation – a case study"; *Mining Technology*; Vol. 112(3); pp. 180-184.

Bonates, E. and Lizotte, Y. (1988); "A computer simulation model to evaluate the effect of dispatching"; *International Journal of Mining, Reclamation and Environment*; Vol. 2(2); pp. 99-104.

Burt, C. N. and Caccetta, L. (2007); "Match factor for heterogeneous truck and loader fleets"; *International Journal of Mining, Reclamation and Environment*; Vol. 21(4); pp. 262-270.

Chironis, N. P. (1985); "Computer monitor and controls all truck shovel operation"; *Coal Age;* Vol. 90(3); pp. 50-53, 55.

Cohen, P. J., Alarie, S. and Gamache, M. (2004); "Multi-stage truck assignment system and method"; U.S Patent issued on May 25th, 2004; Assignee: Caterpillar Inc.

Dessureault, S. D., Scoble, M. and Dunbar, S. (2001); "A study of the synergy of new technology, business tool and information to optimize mine production control"; *International Journal of Mining, Reclamation and Environment*; Vol. 15(1); pp. 6-32.

Gholamnejad, J. and Osanloo, M. (2007); "Using chance constrained binary integer programming in optimizing long term production scheduling for open pit mine design"; *Mining Technology*; Vol. 116(2); pp. 58-66.

Govindaraj, M., (2002); "Strategies for production optimization in open pit mines through simulation models"; Ph. D Thesis submitted to Mangalore University, Karnataka, India (unpublished).

Huang, Y. and Espley, S. (2005); "A 3-D mine simulation model for decision making in mine design and production"; *International Journal of Mining, Reclamation and Environment*; Vol. 19(4); pp. 251-259.

Lewis, M. W., Werner, J. and Sambirsky, B. (2004); "Capturing unrealized capacity"; *CIM Bulletin;* Vol. 97; No. 1076.

Michalakopoulos, T. N. (2001); "Truck dispatching systems"; *Technical Chronical Science Journal,* TCG; Vol. 21(1-2); pp. 37-50.

Nelson, M. G. and Peterson, B. J. (2002); "Simulation of truck/shovel operations with realistic haul roads and 3-D animations"; *Proceedings of the 30th International*

Symposium on the applications of computers and operations research in the mineral industry; Phoenix, AZ; USA; 25th to 27th February; pp. 805-807.

Ramazan, S. and Dimitrakopoulos, R. (2004); "Traditional and new MIP models for production scheduling with in-situ grade variability"; *International Journal of Mining, Reclamation and Environment*; Vol. 18(2); pp. 85-98.

Sturgul, J. R. (1999); "Discrete mine system simulation in the United States"; *International Journal of Mining, Reclamation and Environment*; Vol. 13(2); pp. 37-41.

White, J. W. and Olson, J. P. (1986); "Computer-based dispatching in mines with concurrent operating devices"; *Mining Engineering* (Littleton, Colo); Vol. 38(11); pp. 1045-1053.

In: Characterization and Development of Novel Materials ISBN: 978-1-61324-149-3
Editor: Rafiq Islam

Chapter 4

INTERACTION BETWEEN DATE OF SOWING AND SPACING ON GROWTH AND YIELD OF SENNA (*CASSIA ANGUSTIFOLIA*)

Wasim Aktar[1]*, Sirajul Islam[2] and M. Paramasivam[1]
[1]Deptt. of Agricultural Chemicals,
[2]Deptt. of Agronomy,
Bidhan Chandra Krishi Viswavidyalaya,
Mohanpur, West Bengal, India

ABSTRACT

A field experiment was conducted on Senna (cv – Rabi 20/105) during 2004 to 2005 at Kalyani, Nadia, west Bengal in split plot design in replicate on sandy clay loam soil with eight dates of sowing and two spacing (15 cm × 30 cm and 45 cm × 30 cm), to study their effects on the growth and growth attributes and yield of Senna. The date of sowing and spacing significantly influenced the growth attributes i.e. plant height, dry matter production, crop growth rate, Leaf area index.

Keywords: Senna, Date of sowing, Spacing, Growth attributes, yield.

1. INTRODUCTION

Cassia angustifolia (Family:-Caesalpiniaceae), popularly known as Senna, is a valuable plant drug in ayurvedic and modern system of medicine for the treatment of constipation (Atal et al.,1982; Das et al. ,2003; Maetindale, 1977 and Sharma,2004). Sennoside A and B are the two anthraquinone glycoside that are responsible for purgative action of senna. A variety of poly-herbal formulations containing senna leaves are available in India to relief constipation and allied troubles. Senna is a strong purgative that should be taken in proper

* Correspondence to: wasim04101981@yahoo.co.in.

dosage otherwise it may lead to gripping and colon problem (Bhattacharjee, 2004). Its special morphological characters with respect to deep rooting and drought tolerance have made this crop adaptable in wide range of unfavorable growing conditions with uncertain rainfall and varied soil depth. One of the most significant achievements in the field of grains legume cultivation in India is that Senna can be grown as Rabi crop (Sengupta and Roy, 1982) particularly in the areas where winter is mild and short like Bihar, West Bengal and Andhra Pradesh. Better productivity in rabi season may be achieved due to less attack of wilt and other pests and diseases except pod borer. Hence, more emphasis has to be given under dry land situation on post rainy season Senna for increasing production with simultaneous reduction of area of high water exhaustive crops.

2. Materials and Methods

A field experiment was conducted during 2004 to 2005 at "C' Block farm, B.C.K.V., Kalyani, Nadia, West Bengal. The soil was sandy clay loam soil having organic carbon 0.562%, total N 0.052%, available P 35 Kg ha^{-1} and available K 20.65 Kg ha^{-1} with P^H 7.3. The experiment was laid out in split plot design with 3 replication. The treatments comprised eight (8) dates of sowing 17th (D_1), 25th October (D_2), 2nd (D_3), 9th (D_4), 16th (D_5), 23rd (D_6), 30th November (D_7) and 7th December (D_8) allotted randomly in the main plots and two(2) spacing in the subplots 15 cm × 30 cm (C_1) and 45 cm × 30 cm (C_2) .The treatment combinations consisted of D_1C_1, D_1C_2,D_2C_1, D_2C_2,D_3C_1, D_3C_2,D_4C_1, D_4C_2,D_5C_1, D_4C_2 ,D_6C_1, D_6C_2,D_7C_1, D_7C_2 and D_8C_1, D_8C_2 ,to study their effect on phenological stages and growth attributes of post rainy season Senna (cv – Rabi 20/105).

3. Results and Discussion

3.1. Effect of Date of Sowing and Spacing on Phenological Development of Senna

The number of days for emergence (Table 1) were found to be more or less same for all the sowing dates (3–4 days).

The number of days taken from emergence to secondary branch was maximum in case of late sown crop i.e. D_8 (84 days) followed by D_7 (82 days). The intermediate sowing dates (D_1 to D_6) took less number of days to complete branching. The late sown crop (D_5 to D_8) took more time for flower bud initiation (137 to 138 days) where as earlier sowing (D_1 and D_2) required minimum time for flower bud initiation (124 days).

But in contrary, the late sown crop took more time to attain 50% flowering. Though the earlier sown crop required more time for flower bud initiation, the active reproductive phase in those treatments (i.e. pod initiation to maturity) was very much shortened. Incase of D_6, D_7 and D_8 it was 41, 36 and 37 days respectively as against 64, 76, 54 and 49 days in case of D_1, D_2 and D_3, D_4 respectively.

As a whole the earlier sowing required more time to achieve harvest maturity which progressively decreased with delay in sowing.

Table 1. Effect of date of sowing and spacing on Phenological development of Senna

Date of sowing	Phenological stages (in Days)							
	Sowing to emergence	Emerg. to 3rd sec. branching.	Flr. bud formation	Emerg. to 50% flowering	Flr.bud formation pod initiation	pod initial to 50% formation	1st pod to maturity	Emerg. to harvesting
D_1	3	83	124	119	111	125	64	175
D_2	4	75	124	117	97	126	76	173
D_3	3	72	126	112	112	133	54	166
D_4	4	74	130	116	116	123	49	165
D_5	4	75	137	116	109	130	51	160
D_6	3	76	138	112	112	131	41	153
D_7	3	82	138	110	110	134	36	146
D_8	3	84	137	109	102	133	37	139

3.2. Effect of Date of Sowing and Spacing on Plant Height (cm), Dry Matter Production (g/m^2), Crop Growth Rate CGR (g/m^2/day) and Leaf Area Index (LAI) of Senna

3.2.1. Plant Height (cm) of Senna

At 28 DAS Senna ,the crop sown during 9th October to 23rd October (D_4, D_5 and D_6) plant height was significantly lower than earlier or later sown crops. The highest plant height was recorded form D_2 (25th November) and followed by D_8 (7th December). Similar trend was also followed in 56 DAS. The later sown plant recorded in comparatively higher plant height at 154 DAS. The later sown crops were found to significantly taller than earlier sown crop. As a whole, there was irregularity in the general trend of plant height through out the growing period. Only during the last phase of growth, plant height of late shown crop was found to be higher. This implies the continuation of vegetative growth in case of late sown crop. The spacing effect on plant height, on the other hand was clearly discernible. The close spacing (15 cm× 30 cm) recorded higher plant height than the wider ones. Singh *et al.* (2002) also reported the similar result

3.2.2. Dry Matter Production (g/m^2)

The dry matter production at 28 DAS was highest in case of earlier sown crop i.e. D_1 followed by that of late sown treatment (D_2 to D_8). However D_1 was statistically at par with D_2 and D_3 treatment. Similar trend also observed in later date of sampling. At 56 DAS D_1 and D_2 were statistically significant among themselves and were found comparatively much higher than the rest of the treatments. This wide variation might be due to an appreciable spell of rainfall during these mean periods which could have encouraged the vegetative growth of the earlier sown treatments. This wide different was maintained till 112 DAS. Patel (2000) also reported the same result that earlier sown (30 June) of determinant cultivar (G.T. –100) produced higher dry matter. The dry matter. production was significantly influenced by row spacing in all the date of observation. Close spacing resulted in significantly higher dry matter production from unit area. Due to type of spacing the overall dry matter production increased 30-50% inc case of close spacing. Similar result also observed by Razera *et al.* (2001).

3.2.3. Crop Growth Rate (CGR)

The CGR at initial stage (0 – 28 DAS) was the highest incase of earlier sown crop treatment. However, D_1 was statistically at par with D_2 with respect to CGR. Similar trend was also observed in later dates of sampling, except at 84-112 DAS. At 0 – 28, 28 – 56 and 84 – 112 DAS, D_1 and D_2 were statistically at par with each other and were found comparatively much higher than the rest of the treatments. This wide variation might be due to an appreciable spell of rainfall during these mean period which could have enhanced the vegetative growth of the earlier sown treatments. This wide difference was maintained till 84 – 112 DAS except at 56-84 DAS, where as D_3 and D_7 recorded comparatively higher CGR. The CGR was significant influenced by row spacing in all the dates of sampling, close spacing resulted in significantly higher CGR from unit area. Due to closed spacing the overall CGR increased 33 – 38% as compared to wider spacing except 84 – 112 DAS where the effect of spacing was statistically non-significant.

Table 2. Effect of date of sowing and spacing on plant height (cm), dry matter production (g/m^2), crop growth rate CGR ($g/m^2/day$) andleaf area index (LAI) of Senna

Date of sowing	Plant height (cm)		DM (g/m^2)	CGR ($g/m^2/day$)	LAI
	112 DAS	154(at harvest)	112 DAS	112 DAS	112 DAS
D_1	94.23	118.50	112.48	1.34	4.73
D_2	98.57	124.82	100.15	1.23	4.35
D_3	82.02	116.48	82.20	0.79	3.94
D_4	72.13	113.87	76.85	0.72	3.18
D_5	69.07	117.67	79.30	0.92	2.81
D_6	86.72	145.30	70.13	0.71	2.40
D_7	95.45	136.73	64.45	0.58	2.34
D_8	94.72	138.90	61.33	0.58	2.25
SEm(±)	1.83	2.8316	2.6039	0.1014	0.0374
CD (P=0.05)	5.58	8.6097	7.9175	0.3084	0.1136
Spacing					
C_1	88.30	128.83	90.92	0.89	3.64
C_2	84.91	124.24	70.80	0.83	2.86
SEm(±)	0.96	0.84	0.8661	0.0401	0.0208
CD (P=0.05)	2.89	2.53	2.5965	NS	0.0625

3.2.4. Leaf Area Index (LAI)

The LAI of Senna at 28 DAS was highest incase of earlier sown crop i.e. D_1 which progressively decreased by delay in sowing from D_2 to D_8. Similar trend was also found in later date of sampling. At all dates of observation D_1 and D_2 recorded much higher LAI as compared to later dates of sowing. This wide variation might be due to favorable weather condition at initial stages which could lead to leafy vegetative growth. This wide difference among the treatment was continued till 112 DAS. This is supported by the findings of Patel *et al.* (2000) who were of the view that LAI was greatly influenced by the date of sowing. They observed that SennaCv-GT-100 (determinant growth) sown on earliest date attained the highest leaf area index (LAI) and thus absorbed the maximum amount of photo synthetically active radiation. The LAI was significantly influenced by row spacing in all the dates of observation. Closed spacing resulted in significantly higher leaf area production than that of wider spacing. Due to overall increased leaf area, the maximum leaf area index (3.64 at 112 DAS) was found in close spacing over wide spacing.

CONCLUSION

The date of sowing and spacing significantly influenced the growth attributes i.e. plant height, dry matter production, crop growth rate, Leaf area index. As a whole the earlier sowing required more time to achieve harvest maturity which progressively decreased with delay in sowing. The spacing effect on plant height, on the other hand was clearly discernible. The close spacing (15 cm× 30 cm) recorded higher plant height than the wider ones. Close spacing resulted in significantly higher dry matter production from unit area. Due to type of spacing the overall dry matter production increased 30-50% inc case of close spacing. Due to closed spacing the overall CGR increased 33 – 38% as compared to wider spacing except 84 – 112 DAS where the effect of spacing was statistically non-significant. Due to overall increased leaf area, the maximum leaf area index (3.64 at 112 DAS) was found in close spacing over wide spacing.

ACKNOWLEDGMENT

Authors are very much grateful to Directorate of Research, BCKV, Kalyani for providing instrumental facilities in carrying out this research work.

REFERENCES

Ali, M. and Kumar, S. (2005) Yet to see a breakthrough. *The Hindu survey of Indian Agriculture,* pp. 54-56.

Atal CK, Kapoor BM 1982. *Cultivation and utilization of Medicinal plants*, RRL, Jammu Twai, India. P-8.

Bhattacharjee SK 2004. A hand book of Medicinal plant, Pointer Publisher,Jaipur 74-79.

DasPN, Purohit SS, Sharma AK,Kumar T 2003. *A hand book of Medicinal plants*, Agrobios(India), Jodhpur,118.

Jayanna,M. (2001) Effect of dates of sowing on growth and yield of short duration pigeon pea varieties during summer. *Mysore-Jf-Agric-Sci*, 35 (4): 313-315.

Laxminarayana,P.(2003) Response of rabi redgram to dates of sowing and row spacings. *Annals-Agric-Res*, 24 (1): 187-189.

Maetindale 1977. The extra Pharmacopoeia, 27th ed, the pharmaceutical press, London, 1342-1343.

Patel, N.R., Mehta, A.N. and Shekh, A.M.(2000) Radiation absorption, growth and yield of Pigeon Pea cultivars as influenced by sowing dates. *Experimental-Agric,* 36(3): 291-301.

Razera,L.F., Giomo,G.S. and Nakagawa,J. (2001) Pigeonpea (Cajanus cajan (L.) Millsp.) seed quality from late sowing date and different spacings. *Revista-de-Agric-Piracicaba,* 76 (1): 61-69.

Sengupta, K. and Roy, P.K. (1982) Pulses and Oil seeds in West Bengal, Directorate of Agric., W.B., pp. 87-90.

Sharma R 2004. Agro-Techniques of Medicinal plants, Daya publishing House, Delhi, 176-177.

Singh, J.P. and Srivastava, S.K. (2002) Effect of sowing time and row spacing on the performance of early maturing pigeon pea. *Progressive-Agric*, 2002, 2:96-97.

In: Characterization and Development of Novel Materials
Editor: Rafiq Islam
ISBN: 978-1-61324-149-3

Chapter 5

CONVERSION OF INDIAN CELESTITE TO STRONTIUM CARBONATE

R. K. Tiwary[1], S. P. Narayan[2] and O. P. Pandey[1]

[1]School of Physics and Materials Science, Thapar University, Patiala, India

[2]Advance Materials and Process Research Institute (Regional Research Laboratory), Bhopal, India

ABSTRACT

In the present study celestite ($SrSO_4$), the natural ore of strontium from Trichirapalli district in Tamil Nadu (India) containing impurities like Fe, Ba, Si etc. was chemically treated with HCl/HNO_3 to reduce impurity level. After initial acid purification the mass was treated with charcoals of different grades to obtain SrS from celestite. This was finally converted to $SrCO_3$ by treating SrS with Na_2CO_3 solution. In another effort the leached celestite was directly heated with Na_2CO_3 solution to obtain $SrCO_3$. The percentage conversion of $SrSO_4$ to SrS has been studied with respect to time, temperature, mole ratio etc. The study includes X-ray diffraction study of the raw material celestite, reduced SrS and produced $SrCO_3$ for the identification of phases. The SEM study was also conducted for celestite and strontium carbonate for the study of its particle morphology. The details of conversion and the obtained results have been described in this paper. Focus has been also made to optimize the process to prepare $SrCO_3$ (ferrite grade) with maximum purity up to 99.0% $SrCO_3$. The present study also gives a comparative study of the two processes involved in the preparation of $SrCO_3$ in terms of process easiness, economy, percentage yield etc.

Keywords: Celestite, Strontium sulphate, Strontium carbonate, Black ash process, Direct conversion process.

INTRODUCTION

The Mineral Celestite existing in different places across the world is main source for the production of $SrCO_3$, which contains 65 to 90 percent of strontium sulphate [1, 2]. In India it is found in Trichirrapalli district of Tamilnadu state. This is medium grade celestite as it contains large amount of impurities like Fe, Ba etc [18]. Many other strontium chemicals like strontium nitrate, strontium chloride, strontium hydroxide and strontium oxide is being manufactured from celestite [3-7]. The major consumption of strontium carbonate is in glasses for colour television picture tubes to block X-ray transmission and improvement of glass appearance. The next major application of $SrCO_3$ of ferrite grade is in ceramic industries to produce ferrites of different grades for electrical and electronic appliances. Apart from this strontium carbonate is also used in purification of zinc electrolyte (to remove lead impurities) and as ceramic paints [1]. It is also used in farm chemicals, medicine and as tanning material [3].

The Beneficiation, purification of strontium carbonate from the natural ore celestite finds great difficulties to obtain required level of purities in the final product. Black ash method and direct conversion method are the two methods by which conversion of celestite to strontium carbonate is being carried out [4, 5]. A group of researchers have attempted to study the thermodynamic and kinetic parameters of the direct reduction and black ash processes [6, 8]. Apart from this scientists have conducted experiments on conversion of celestite to strontium carbonate using different technique [9-12]. Hydrogen and mixture of H_2 and CO was used at high temperature for the reduction of celestite by Pedak et al. [13]. The major disadvantage of this process is consumption of large amount of gases and evolution of highly pollutant compound like H_2S, which is not eco- friendly.

Gits et al. [14], Lepsin et al. [15] and Erdemoghi et al. [4] have also studied the thermo chemical reaction with a stoichiometric amount of carbon at a temperature 1200-1300°C. The loading zone temperature of rotary kiln was 600-700°C and discharge zone temperature was 1200°C to 1300°C. This process has many disadvantages such as high electrical energy requirements, sintering of reactants and product etc. To overcome such type of problems experiments have been carried out with easily available reductants like coke, bituminous coal, and activated charcoal [5, 6]. There is no data available for the comparative study of direct reduction process and black ash process through simultaneous experiments with respect to percentage yield, application etc. of Indian celestite. The present study aims to prepare strontium carbonate of ferrite grade, which consists of many purification steps. These include black ash method and direct reduction method for the comparative study of conversion of celestite to $SrCO_3$ with respect to process cost, process simplicity, by product application etc. The basic aim of the project is to prepare strontium hexaferrite sintered magnets using celestite ore. In this first report the experiments performed for conversion of Indian celestite to strontium carbonate by different processes is reported.

EXPERIMENTAL PROCEDURE

The celestite ore obtained from Trichirapalli deposit of Tamil Nadu (India) was supplied by M/S J.M. Mines and Minerals, 37, Williams Road Cantonment, Tamil Nadu (India). The ore was ground to fine powder in a mortar pestle. The powder was sieved. The fraction less than 45 micron was collected for the experimental work. The chemical analysis of celestite was carried out to know its chemical composition. For different trace element analysis and atomic absorption spectrometer flame type (Model GBC-932AA, Australia) was used.

Chemical treatment of celestite ore: 100 gm celestite powder (<45µn) was taken in 2 Liter borisilicate glass beaker. About 500 ml of 1:1 HCl, 20 ml conc HNO_3 and small amount of NH_4 F was added in this fine powder mass. All acids and other chemicals were of L.R. grade purity. The beaker containing acid and celestite was heated on a hotplate at a temperature of 60°C with constant stirring for time period of 36 hours. The acid portion was decanted and fresh HCl (1:1) was added followed by small amount of HNO_3 addition to repeat this process. Acid treated celestite was filtered and washed many times with distilled water till the pH of residue on the filter paper attains neutral value. The filter paper with residue was dried in an oven at 110°C for 2 hours. By addition of HCl, HNO_3, the strontium carbonate present with celestite along with Fe, Ca, dissolves and remains in the dissolved condition in the liquid. NH_4F has been added to dissociate any silicate materials present in celestite.

Table 1. Chemical analysis of celestite ore before and after acid treatment

Composition	Before Acid leaching	After acid leaching
$SrSO_4$	91.60	96.66
$SrCO_3$	1.30	<0.02
$BaSO_4$	0.90	0.87
Fe_2O_3	4.01	0.60
SiO_2	0.10	0.10
Al_2O_3	0.21	0.15
CaO	0.36	0.01

Table 2. AAS Results for other trace elements before and after acid treatment

Element	Mg	Pb	Zn	Ni	Cu	Na	K
Process							
(i) Before leaching	0.12	0.24	0.01	0.006	0.29	0.24	0.026
(ii) After leaching	0.012	0.021	0.01	0.005	0.002	0.002	0.021

Table 3. Chemical composition (wt. %) of different charcoals

Coal	Fix carbon	Ash	Volatile material	Moisture
Active Charcoal	97.0	1.7	0.5	0.7
Coke	82.86	12.26	0.2	3.85
Bituminous coal	62.26	25.62	6.32	8.62

By chemical treatment process celestite up gradation is performed and acid soluble impurities like Fe, Ca, etc. reduce significantly. The chemical composition of celestite before and after acid treatment is given in table 1 and 2.

Black Ash Method

In this process celestite obtained after acid treatment is reduced by carbon, at higher temperature. Three different sources of carbon viz. activated charcoal, bituminous coal and coke had been used as the reducing agent. Chemical composition of these is given in table3. The particle size of all three carbons was less than 100 micron.

The reaction involved in this reduction reaction can be written as below.

$$SrSO_4 + 4\,CO \rightarrow SrS + 4\,CO_2$$

$$CO_2 + C \rightarrow 2CO$$

$$SrSO_4 + CO \rightarrow SrO + SO_2 + CO_2$$

The carbon monoxide generated in this process diffuses and reacts with celestite which is not in contact with carbon. The excess carbon used in the process will give rise to formation of CO which will help in gas solid diffusion reaction. To start this experiment treated celestite power (< 45 μm) has been taken and mixed with charcoal (heated at 110°C for 24 hours before mixing). These experiments were done with different amount of charcoal, which varies with 10%, 20%, 30%, 50% and 60% in excess amount (weight ratio) of celestite taken. For example for 100gms celestite 110 gm of charcoal was taken and similarly the amount of charcoal was varied. This mixture was kept in silica crucible and covered with silica cover lid. Thick ceramic paste was applied around the cover. The system was heated in a muffle furnace at 1000°C for 100 minutes. This experiment was conducted with different grades of coal at the same temperature with variation in reducing time. The reduced SrS was leached with water. A little amount of HCl was added to dissolve non-soluble SrS. An excess amount of H_2SO_4 (6-10 times more then theoretical amount) was added to remove Ba as $BaSO_4$ and for it a long reaction time period was employed (8 hours) [5].

The leached solution was filtered through the qualitative filter paper. This facilitates the removal of silicate and other insoluble constituents. The dissolved portion has been used for the precipitation of strontium carbonate. A 10% solution of sodium carbonate was added to the dissolved acidic strontium solution with constant stirring at 50°C for precipitation of strontium carbonate. The precipitated strontium carbonate was filtered through qualitative filter paper and washed with distilled water many times until the residue on the filter becomes neutral pH.

Direct Conversion of Celestite to Strontium Carbonate

In another effort celestite after leaching treatment was treated with Na_2CO_3 solution. The reaction involved in this process can be written as:

$$SrSO_4 + Na_2CO_3 \rightarrow SrCO_3 + Na_2SO_4$$

For this purpose 100 gm powder (<45μm) of treated celestite and sodium carbonate was taken in mole ratio of (Na_2CO_3 / $SrSO_4$) 1.0, 1.1, 1.2, 1.5. The sodium carbonate in required quantity was dissolved in 1 liter solution. The celestite powder was added slowly to the carbonate solution under the constant stirring at a temperature ranging from 25°C to 75°C at an interval of 10°C with time variation from 30 minutes to 330 minutes at an interval of 30 minutes. The beaker containing solution was cooled and the converted $SrCO_3$ was filtered through the filter paper. The sodium sulphate produced in this reaction remains in the dissolved form. The residue on the filter paper was washed many times with distilled water until the residue becomes neutral.

The filtered and dried mass was again taken in 2 liter borosilicate beaker and dissolved in (1:1) HCl at a temperature of 40°C for 1 hour. After cooling the beaker the liquid in the beaker was filtered through filter paper. In this way any silicate material and un-dissolved impurities was collected on filter paper leaving Sr as $SrCl_2$ in the solution. It was observed that Na_2CO_3 /$SrSO_4$ of molar ratio 1.5 give better yield. $SrCO_3$ obtained from 1.5 molar ratios (Na_2CO_3/$SrSO_4$) was again added to sodium carbonate solution with constant stirring to precipitate strontium carbonate. The $SrCO_3$ obtained was analyzed by wet chemical analysis.

The study of phase analysis for the raw material celestite and the product SrS, $SrCO_3$ formed during reduction of celestite was conducted by XRD (Model Rigaku D-Max IIIC) and particle morphology was analyzed by scanning electron microscope(Model JEOL 840A).

Results and Discussion

Black Ash Method

Celestite ore consists of $SrSO_4$ as major constituent as can be seen from X-ray diffraction pattern (Figure 1). The reduction of celestite was done with different grades of coal. In all the experiments excess amount of charcoal by weight percentage was taken. The celestite to charcoal ratio was varied as 1:1.2, 1:1.3, 1:1.4 and 1:1.6.

Figure 2 shows the effect of excess activated charcoal content on reduction of celestite with variation in reducing time at 1000°C. The reducing effect decreases considerably when the activated charcoal is less than 40% excess of the celestite mass. The reducing effect of activated charcoal with 60% excess mass was better. The conversion process is completed after 60 minutes of reduction time after that it becomes virtually constant. As the percentage of excess charcoal with respect to celestite mass increases the conversion percentage also increased.

From the figure it can be observed that up to a time period of one hour conversion increases considerably with each coal percentage mass. After one hour the conversion percentage rate becomes relatively slow. Similar trends were also observed for bituminous coal and coke, which are shown in figure 3 and figure 4 respectively with slight change in rate of conversion. Figure 5 shows the typical X-ray diffraction pattern of reduced celestite where peaks of SrS have been marked.

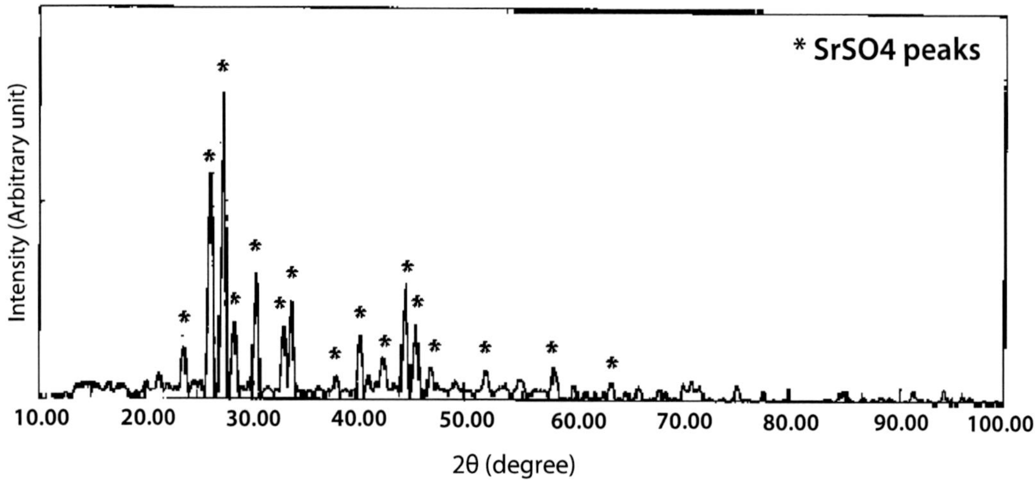

Figure 1. Typical X-ray diffractogram of celestite showing the peaks of SrSO4.

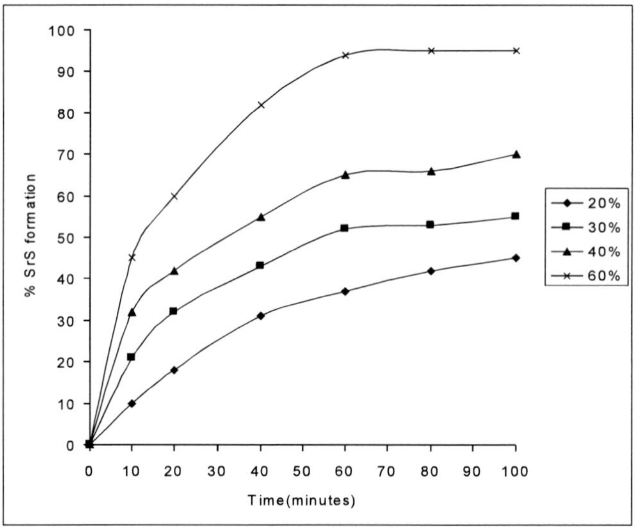

Figure 2. Effect of reducing time and excess activated charcoal on % yield of SrS at 1000oC.

The work by Wen Chan [5] for the reduction of Chinese celestite with different carbon source like brown coal, bituminous and coke indicates that the reducing effect with 40% excess charcoal with mass ratio of celestite is minimum and with 60% of mass ratio percentage conversion of celestite to SrS is maximum. The impurities associated with celestite such as Ba, Fe, Ca and Al has been separated in a sequence using chemical routes. However, the Chinese (Lamping) celestite contains around 4.0-8.0 % silica as impurities. It is mentioned that silica will react with the Sr present in the celestite and form a complex oxide $SrSiO_3$ that is acid soluble salt for that constant pH has to be maintained for longer duration. But it is not necessary that all silica present in the celestite may react with celestite.

Moreover, the purity of the product is limited to 92%. Therefore it may be possible that some residual amount of silica may remain as contaminated phase with final product. Since the silica content in the present study is very low the possibility of silica contamination is very rare.

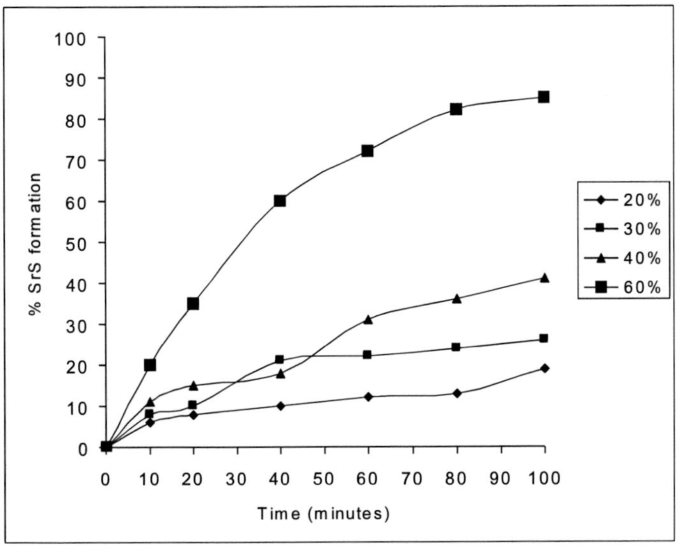

Figure 3. Effect of reducing time and % excess bituminous coal on % yield of SrS at 1000oC from celestite.

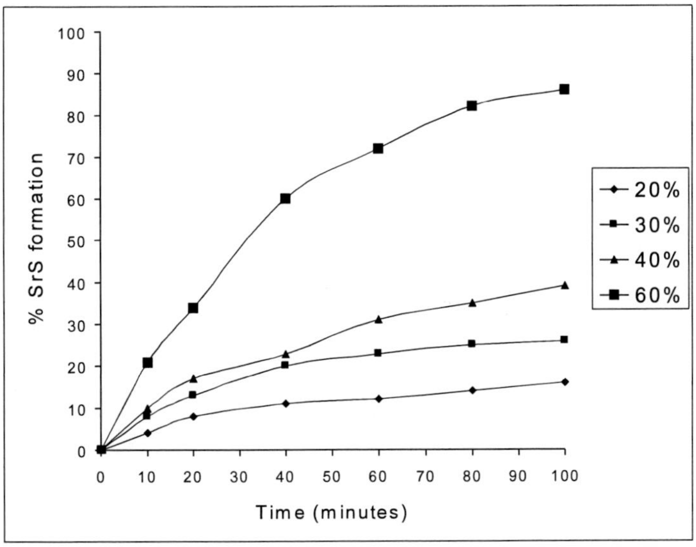

Figure 4. Effect of reducing time and % excess of coke on % yield of SrS at 10000C.

The study of Sonawane et al. [6] shows that at a mole ratio of 2.5 (activated charcoal to celestite) the reduction of celestite into strontium sulphide increases with increase in temperature and time. The kinetics of the reduction reaction has been studied using different catalyst like potassium carbonate, sodium carbonate etc. The reduction temperature has reduced substantially. His study was confined to see the effect of catalyst on reduction temperature. The removal of catalyst from the final product is not specified in this work, which is around 2%. Gits et al. [14], Lapsing et al. [15] and Erdemaglu et al. [4] have also shown the similar type of reduction behavior at higher temperature with different time period. The disadvantage of these experiments is that this reaction is conducted at very high temperature 1200-1300^{0}C consuming lot of electrical energy. Recently Abdullah Obut [16] has reported an experiment for the conversion of celestite to strontium sulphide by microwave heating. The conversion ratio in this experiment is observed to be 98%. But the temperature of conversion was 1200^{0}C.

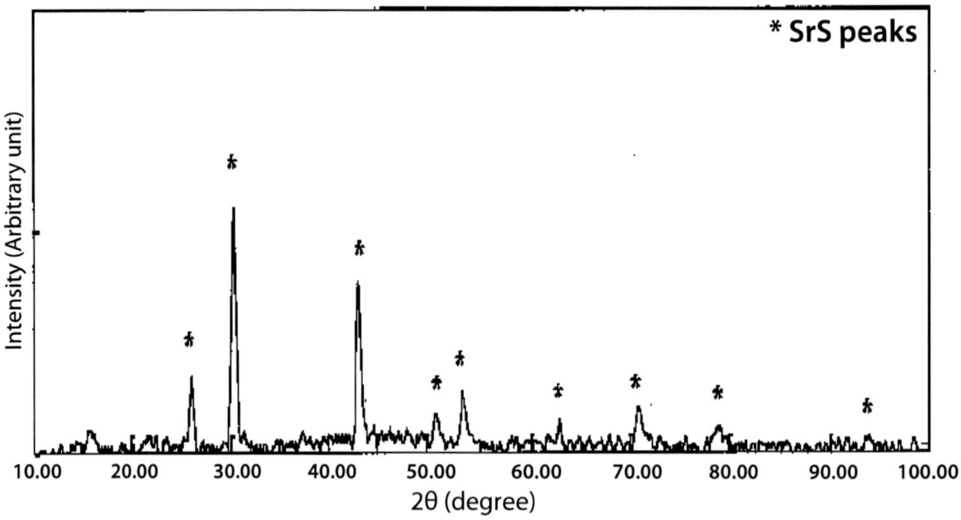

Figure 5. Typical X-ray diffractogram showing the peaks of SrS obtained from black ash process.

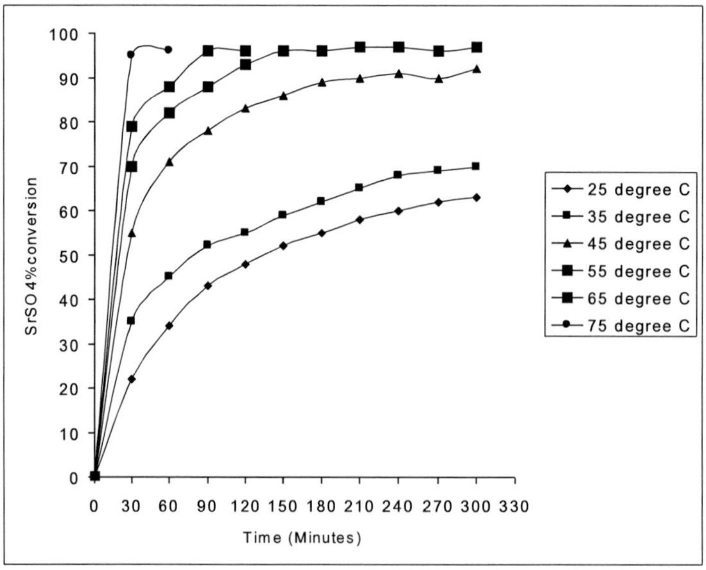

Figure 6. Effect of temperature and time on % yield of SrCO3 for mole ratio 1.5 (Na2CO3/SrSO4).

In our study the results obtained with activated charcoal is better as compared to other grades of coal. The reducing time is also less as compared to other reported work [5].

SrS obtained after reducing experiments was leached in acidic solution. The reaction involving in the leaching process of strontium sulphide can be written as

$$2\ SrS(s) + 2H_2O \rightarrow Sr\ (HS)_2 + Sr(OH)_2$$

$$Sr(HS)_2 + 2H_2O \rightarrow Sr\ (OH)_2 + 2\ H_2S$$

$$SrO(s) + H_2O \rightarrow Sr\ (OH)_2$$

$$Sr(HS)_2 + 2HCl \rightarrow SrCl_2 + 2H_2S$$

H_2SO_4 added to solution will precipitate Ba as $BaSO_4$. About 70-75 percentage Ba^+ gets precipitated where Sr^+ as strontium sulphate loss percentage is 0.98%. To remove Ba^+, Ca^+ up to 95% the pH of the solution was increased to 12 [5]. The Ca^{+2}, Mg^{+2}, Fe^{+2} ions hydrolyze at pH 12. So up to 95% of these impurities are eliminated at pH 12 due to hydrolysis.

Direct Conversion Method

Figure 7 shows the effect of temperature and time on percentage yield of $SrCO_3$ for mole ratio 1.5 (Na_2CO_3/ $SrSO_4$). The experiment carried out at 25^0C exhibits the lower percentage of celestite conversion. As reaction temperature increases from 25 to 75^0C the celestite conversion to strontium carbonate increases rapidly.

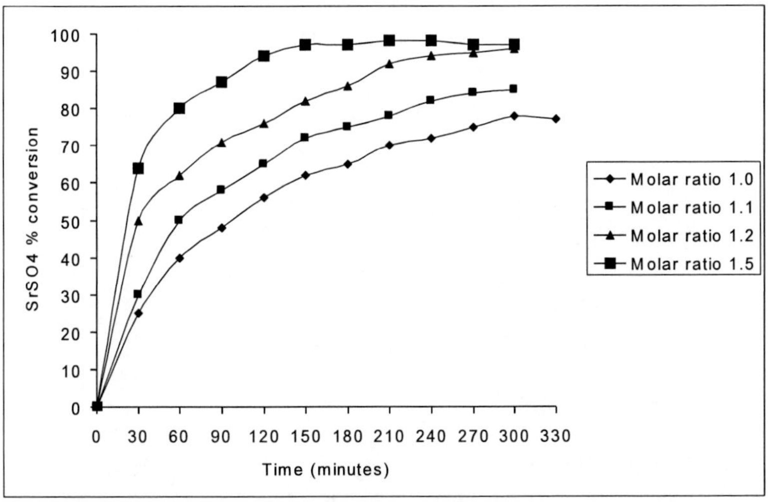

Figure 7. Effect of time and mole ratio (Na_2CO_3/$SrSO_4$) on % yield of $SrCO_3$ from celestite at 60^0C.

The effect of time and mole ratio (Na_2CO_3 / $SrSO_4$) on percentage yield of $SrCO_3$ from celestite at 60^0C has been shown in figure 8. For molar ratio of 1.0 in the present study, the percent conversion is quite low. The conversion rate of celestite increases with increase in molar ratio. For molar ratio 1.5 the conversion rate is maximum. Castillejos et al [9] performed experiments with different mole ratio (Na_2CO_3 / $SrSO_4$) 1.0 to 1.5 at 45^{0}C for a time period of 350 minutes. He reported the sustained increase in the rate of reaction with an increase in sodium carbonate concentration. The chemical composition of celestite as given in table-1 shows the presence of large amount of Fe (4.01%) which after acid leaching reduced to 0.60% as shown in the table 1. The percentage of Al_2O_3 content also reduced from 0.2 to 0.15%. In this acid leaching of celestite powder impurities like Fe, Al and Ca etc. has been reduced to great extent. Leaching also helps in removing other soluble impurities. The celestite leaching experiment employed by Hacer Dogan et al [7] led to the reduction of impurities up to 0.005%Fe, 0.11%Ca 0.6%Ba, 0.86%Pb and 0.2% others.

The level of impurity reduction in this experiment seems to be in good agreement with value reported by Hacer Dogen et al. [7]. The reduction of celestite with different coal shows that the best result was obtained with the activated charcoal in terms of percentage yield of SrS with respect to time as shown in figures (3,4 and 5). The reducing property decreases considerably as the excess carbon decreases from 60% to 40% of sample mass of celestite. The conversion percentage increases as time increases with each of the coal sample. The SrS conversion conducted by Wen Chen [5] using various source of coal indicated that brown coal gives maximum efficiency for the conversion from celestite to SrS. However, the present study shows the activated charcoal is most efficient coal source, which can be used in reduction of celestite to strontium sulphide. The difference in the result is due to the fact that Wen Chan [5] has not employed the activated charcoal, which is rich in fix carbon containing lower value of ash. Whereas, the study made by Sonawane et al.[6] for the reduction of celestite into SrS reports that the activated charcoal is most effective reducing substance for celestite reduction. In our study also we have found the similar trend. The effect of different molar ratio 1.0, 1.1, 1.2, 1.5 of (Na_2CO_3 /$SrS0_4$) on conversion of $SrCO_3$ with respect to time (ranging from 30-330 min. at 45°C has been shown in figure (7).

Figure 8. SEM micrograph of strontium carbonate.

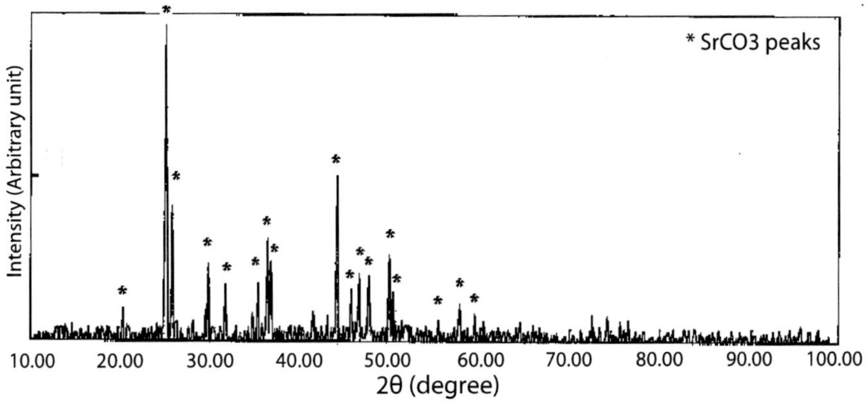

Figure 9. X-ray diffractogram showing the peaks of strontium carbonate obtained from black ash method.

The rate of percentage conversion of SrS is found to be maximum with 1.5 molar ratios of (Na_2CO_3 / $SrSO_4$). The decrease in the molar ratio exhibits the lower percentage of SrS with respect to time. Castillejos et al. [9] have reported that at a mole ratio (Na_2CO_3/ $SrSO_4$) of 1.5 maximum conversion rate of celestite to strontium sulphide was observed which matches with the result of the present study.

Purification process to get strontium carbonate of desired level of purity by black ash and direct decomposition process after removing impurities is up to 97 and 98 % respectively.

Figure 8 shows the SEM micrograph of $SrCO_3$ where powders of different shape and size can be seen. The variation in particle size is from 2 μm to 10 μm.These particle get agglomerated. A careful examination indicated that smaller size powders are spherical in shape where as bigger one is faceted type. The X-ray diffraction pattern of these powders shows the presence of single phase $SrCO_3$ compound (figure 9).

Comparison of Black Ash and Direct Conversion Method

Table-4 shows the chemical composition obtained from both processes. The yield percentage is greater in direct conversion method (98%) as compared to black ash method (97%). Kirk-othemar, (Encyclopedia) [1] has reported a yield of 95% strontium carbonate from black ash process. According to a Patent [17] it has been mentioned that it is possible to produce 95% pure strontium carbonate by using black ash method.

However, this requires the celestite containing 85% strontium sulphate whereas by using direct conversion method strontium carbonate over 95% can be achieved. It was also claimed that pure strontium carbonate (99%+pure) can be also obtained even with low-grade celestite by double decomposition method [17]. The purity of strontium carbonate obtained in present study seems to be match well within the limit of the values reported by other workers [4,5and6]. By comparison of both processes it seems that direct decomposition process is more economical as compared to black ash process as no extra equipment like furnace is required. Only the chemicals are required. Moreover, the by-product obtained can also be utilized where as it is not possible in black ash process.

Table 4. Comparison of the process in terms of % yield and residual impurities

Process	$SrCO_3$	$BaCO_3$	$(Ca^{+2} + Mg^{+2})$	Fe	Si
Direct decomposition method	98.01	< 0.01	0.01	0.008	0.1
Black ash method	97.85	<0.01	0.015	0.01	0.02

CONCLUSION

The reduction of celestite to strontium sulphide depends upon the different variables like quality of reducing agents, temperature and time. The activated charcoal was found most suitable for reducing celestite. The other coals like bituminous coal and coke render low percentage of conversion of SrS. In the direct conversion method conversion of celestite is also dependent of various parameters like molar ratio ($Na_2CO_3/SrSO_4$) temperature time etc. We can achieve maximum yield of $SrCO_3$ by optimizing the various parameters. By comparison of black ash and direct conversion method it appears that the latter one is better process since it does not consist of big set up like reducing gaseous set up as used in black ash process. Direct reduction method has the advantage of simplicity, low consumption of acidic material and high recovery of strontium compound. The strontium carbonate produced from this process is of high purity, which can be used in the production of ceramic ferrite magnets. The direct conversion method may be suitable for high-grade celestite where Ba like impurities is less. The single decomposition method may be suitable for low grade strontium carbonate where as double decomposition method may be most suitable for ferrite grade strontium carbonate replacing the expensive pure $SrCO_3$.The Indian celestite contains medium amount of impurities like Fe, Ba etc. and therefore may be graded as medium grade celestite. Before conversion of celestite by direct decomposition method it is necessary to perform celestite up gradation by acid leaching to remove the acid soluble impurities. The Indian requirement of strontium carbonate is almost furnished by import from other countries. India has a large deposit of celestite. There is need to explore the indigenous know how for better use of celestite to start production of strontium carbonate at large scale. This will help the country to utilize the celestite ore and make them self-dependent as well as save the foreign money in turns

ACKNOWLEDGMENT

The authors are thankful to Department of Mines, New Delhi for providing financial assistance for this work.

REFERENCES

[1] Krik-othmer,(1993) "*Encyclopedia of chemical technology*", fourth edition 10, pp 393-405, John Willey and Sons.

[2] Joyce A. Ober.,(2007) *Strontium USGS Report.*

[3] J Pual, Mac Millan,(1994) "Strontium and Strontium compound", *Ullmann's Encyclopedia of Industrial Chemistry*, A-25.

[4] M.Erdmoghu, M.Canbazughu, H.Yalcin,(1998) "Carbothermic reduction of high-grade celestite ore to manufacture strontium carbonate". *Transaction of the Institution of Mining of Metallurgy, See-C, Mineral Processing and Extractive Metallurgy*, vol. 107, pp.65-68.

[5] W.Chen, Y.Zhu,(2000) "Preparation of strontium carbonate from celestite" *Mineral Processing and Extractive Metallurgy IMM Transaction C*, , vol 109, pp. 1-5.

[6] R.S. Sonawane, B.B.Kale, S.K.Apte and M.K. Dongare, (2000) "Effect of catalyst on the kinetics of reduction of celestite ($SrSO_4$) by active charcoal", *Metallurgical and Materials Transaction B*, Vol-31B, pp 35-41.

[7] Hacer Dogan, Murat Koral, Sidika Kocakusak, (2004), "Acid leaching of Turkish celestite concentrates", *Hydrometallurgy* vol.71, PP 379-383.

[8] Bronislaw Janczuk, Emil Chibowski, Tomasz Biajopiotrowicz and F.Gonzalez-Caballero, (1989) "Surface free energy of celestite and its floating activity", *Collides and Surfaces*, vol-35, issue -1, pp 41-48.

[9] A.H. Castillejos E, F.P. de la Cruzdel B, A. Urbis S.(1996), "The direct conversion of celestite to strontium carbonate in sodium carbonates aqueous media", *Hydrometallurgy* ,vol. 40 pp.207-222.

[10] Massone, J.,(1983), "Technology and uses of barium and strontium carbonate." In: B.M.Coope and G.M. Charke (editors) *5th industrial Minerals int.Cengr. (Madrid, Spain, 1982) Metal Bulletin, London*, pp 115-119.

[11] Sutarno, R.H Lake and W.S Bowman,.,(1970), "The extraction of strontium from the mineral *Mines and Resource's Ottawa"*, Res. Rep.(R-223), pp. 1-20.

[12] M. Iwai. and J.M.Toguri, (1989), "The leaching of celestite in sodium carbonate, *Hydrometallurgy"*, vol. 22, pp. 87-100.

[13] E. Pedak, M. Allsalu and M. Kanter Zh. (1972), "*Prikl Khim*, (*Leningrad*)", vol. 45(12), pp. 2619-23 (In Russian).

[14] E.B.Gitis, P.I.Strigunov, S.K.Solyanik, L.F.Vasileva, F.D. Pushkarnov, Y.N. Zinchenko and V.I. Krasnyakove:(1985), "*Otkritiya Izobret"*, vol.-46, pg. 91.

[15] V. A. Lepsin, L.E. Kruglove, G.R. Bulgakove and V.I. Krashyakov:(1988), "*Otkrytiya Izobret'*, vol. 10, p. 89 (In Russian).

[16] Abdullah Obut,(2007), "Direct conversion of celestine to SrS by microwave heating," *Minerals Engineering",* (Article in press).

[17] de Buda,(1988), *United States Patent* No- 4,666,688, pp 1-8.

[18] www.tidco.com/image%5cstrontium%20carbonate.doc

In: Characterization and Development of Novel Materials
Editor: Rafiq Islam
ISBN: 978-1-61324-149-3

Chapter 6

CHARACTERIZATION OF CLINICAL BACTERIA AND THEIR CONTROL BY A MEDICINAL PLANT (*CASSIA SOPHERA L.*) EXTRACTS

***M. Mizanur Rahman*[1], *M. Mashiar Rahman*[2], *Rezuanul Islam*[2], *Atiqur Rahman*[3] *and M. Safiur Rahman*[4*]**

[1]Dipartimento di Biotecnologie Agrarie, Viale dell'Università 16, Legnaro (Padova), Italy
[2]Department of Biotechnology and Genetic Engineering, Islamic University, Kushtia, Bangladesh
[3]Department of Biotechnology, Daegu University, Kyoungsan, Korea
[4]Environmental Analytical Chemistry Lab, NRCD, Institute of Nuclear Science and Technology, Bangladesh Atomic Energy Commission, Dhaka, Bangladesh

ABSTRACT

The effectiveness of a medicinal plant (*Cassia sophera* L.) for controlling the growth and survival of microorganisms isolated from stool samples collected from Kushtia General Hospital, Bangladesh was studied. Biochemical tests identified some bacteria such as *Staphylococcus aureus*, *Enterococcus species*, *Pseudomonas aeruginosa, Escherichia coli* and *Klebsiella pneumoniae*. The antibacterial activity of various organic solvent (petroleum ether, chloroform and ethanol) extracts of *C. sophera* leaves was determined *in vitro* using agar diffusion method and MIC against the isolated bacteria. The extracts (300 μg $disc^{-1}$) displayed potential of antibacterial activity against two Gram-positive bacteria (*S. aureus* and *Enterococcus* sp.) and two Gram-negative bacteria (*Klebsiella pneumonia* and *E. coli*). The zones of inhibition of different organic solvent

[*] Corresponding author: Department of Environmental Engineering, Faculty of Engineering, Dalhousie University, D215-1360 Barrington St., Halifax, N.S. B3J 1Z1, Canada. Email: Safiur.Rahman@dal.ca.

extracts against the tested bacteria were found in the range of 12.1 ± 0.5 to 21.3 ± 1.1 mm and the MIC values were recorded between 62.5 and 500 μg ml^{-1}. The antibacterial activity was compared with commercial antibiotics such as streptomycin and tetracycline. In most of the cases, the extracts (chloroform and ethanol) exhibited similar or higher antibacterial activity than standard drug streptomycin. The results of this study suggest that the extracts of *Cassia sophera* L. can be a source of natural antimicrobial agents with potential applications in pharmaceutical industries to control pathogenic bacteria causing severe illness in humans.

Keywords: *Cassia sophera* L., pathogenic bacteria, antibacterial activity, MIC,

1. INTRODUCTION

Infectious diseases caused by pathogenic microorganisms are a major health concern. Intestinal infection is the most common cause of diarrhea worldwide through enteric bacteria and is estimated to be responsible for the deaths of 3–4 million individuals each year, particularly infants and young children [WHO,1996]. Acute watery diarrhea accounts for approximately 80% of such episodes, persistent diarrhea for 10%, and dysentery for up to 10% [Bhan, 2000]. Diarrhea is a disease commonly encountered both in urban and rural centers of Bangladesh. Risk factors that predispose children to diarrhea include poor sanitation, poor social economic status and malnutrition [Njoku-Obi and Anozie, 1984]. Diarrhea remains public health problem all over the world and acute diarrhea due to bacterial infections is one of the main causes of childhood morbidity and mortality, especially in developing countries [Katouli, et al 1990].

The major burden of infection is due to food-borne infections and water-borne infections caused by *Staphylococcus aureus*, *Escherichia coli*, *Enterococcus species*, *Pseudomonas aeruginosa*, and *Klebsiella pneumoniae*. Antibiotics are used in treatment of some bacterial infections. The increasing use of antibiotics and misuse by over-prescribing and/or poor patient compliance has led to the development of bacterial resistance to antibiotics [Kotzé and Eloff, 2002]. This resistance has catalyzed the search for new antibacterial agents from natural sources such as extracts from plants [Cowan,. 1999, Nasar-Abbas and Halkman, 2004]. The use of a wide range of plant extracts and other natural substances to enhance animal health, growth and performance has been advocated for a long time due to their broad pharmacological activities [Conner, 1993, Rhodes, 1996]. The presence of antibacterial activity in plants is important from an ecological as well as pharmacological viewpoint [Cox and Balick, 1994].

Cassia sophera L. (Caesalpiniaceous) is a medicinal plant of Bangladesh and Indian sub continent, which is widely used as antihistamine and antioxidant. According to the physicians of Unani medicine, three plants viz., *C. occidentalis* Linn., *C. sophera* Linn. and *C. sophera* Linn. var. *purpurea* Roxb. are the varieties of 'Kasondi' and are invariably conditions [Ahmad-Billal,et al 2005, Awan, 1984, Lubhaya, 1975]. In ethno botanical literature, it is mentioned to be effective in the treatment of pityriasis, psoriasis, asthma, acute bronchitis, cough, diabetes and convulsions of children [Agharkar, 1991, Chopra et al., 1956, Drury, 1873, Dutt, 1995, Dymock,et al 1990, Kirtikar and Basu, 2000, Nadkarni, 1982].

Owing to its various ethnopharmacological properties, no report available in the literature on the screening of different solvent extracts of *C. sophera* leaves for their antibacterial properties against some clinical enteric bacteria. Hence, efforts have been made here, specifically to investigate the role of various solvent extracts of *C. sophera* as a potential antimicrobial agent against some enteric bacteria causing infectious diseases such as diarrhea and dysentery.

2. Materials and Methods

2.1. Collection of Stool Samples for Isolating Bacteria

A total of 100 stool samples were collected from children ages between 0 – 10 years. Thirty of the samples were collected from hospitalized children in the Kushtia General Hospital with the age group ranging from less than 6 months to 5 years. Ten samples each were collected from seven public primary schools. The samples were collected in sterile specimen bottles having the names and ages of patients written on the sides of each bottles. Care was taken to select pupils who have not received treatment with antibiotics for four weeks prior to sample collection. All the samples were transferred immediately into the laboratory for analyses. Selective and differential solid media as well as enrichment broth were used for the primary isolation of enteric organisms (Table 1). Those include MacConkey agar, deoxychocolate citrate agar (oxoid) and selenite F (oxoid), an enrichment broth.

2.2. Biochemical Tests for the Characterization of Bacteria

Biochemical tests such as Gram Stain, Oxidase Test, Catalase Test, Indol Test, Citrate Test, Mannitol Test and Litmus Milk Decolorization Test were carried out on the bacterial isolates using conventional techniques, as described earlier [Cowan, 1999].

2.2.1. Gram Stain

Using a sterile loop, a loopful of tap water was transferred to a clean glass slide. The desired colony was touched by a loop and the specimen was allowed to dry on the slide at room temperature. After, the specimen dried and fixed by gently heated the slide by passing quickly through the flame (3-4 times). The bacterial smear was stained by the primary stain crystal violet, incubated for 1 minute and washed gently by tap water for 2-3 seconds. Then, iodine (Gram's iodine) solution (1% iodine, 2% potassium iodide in water) was added for 1 minute. This acted as a mordant and the dye was fixed. It was rinsed with tap water and 95% ethanol was applied for 10 seconds. To prevent over-decolorisation, it was rinsed immediately with water. Finally, the secondary stain safranin was added and incubated for 1min, and then washed with water for a maximum of 5 seconds. The slide was observed under a microscope.

2.2.2. Oxidase Test

The Oxidase test was used to assist in the identification of *Pseudomonas*. To perform this, a piece of filter paper was placed in a clean petridish and 2 or 3 drops of freshly prepared oxidase reagent was added. The colony of test organism on the filter paper was smeared.

2.2.3. Catalase Test

This test was used to differentiate those bacteria that produced the enzyme catalase, such as Staphylococci from non-catalase producing bacteria such as Streptococci. The hydrogen peroxide solution (2-3 ml) was poured into a test tube. Using a sterile wooden stick or glass rod, several colonies of the test organism were removed and immersed in the hydrogen peroxide solution and the development of an immediate froth of bubbles indicated a positive catalase test. The test was performed on a blood-free medium.

2.2.4. Indol Test

The test organisms were cultured in tryptophan containing medium in a bijou bottle with 3 ml of sterile tryptone water. The culture medium was incubated at 35-37°C for up to 48 h. 0.5 ml of Kovac's reagent (4(p)-dimethylamino-benzaldehyde) was added in the medium with gentle shaking and observed for bright pink color in the surface layer within 10 minutes for only smooth pink bacterial colony.

2.2.5. Citrate Test

For detection of particular type of bacteria, a dense bacterial suspension of the test organism in 0.25 ml sterile physiological saline was inoculated in small tube. The tube was incubated overnight at 35-37°C after adding citrate tablet. Color changes were observed. Red color incase of smooth pink and pale colored colonies indicating positive citrate test, where yellow orange color for other bacterial colonies indicating negative citrate test.

2.2.6. Mannitol Test

Mannitol salt agar is a selective and differential medium used for differentiating between different staphylococci. *S. aureus* changes medium to yellow, whereas *S. epidermidis* will not change the medium.

2.2.7. Litmus Milk Decolorization Test

Using a sterile loop, 5 ml of sterile litmus milk medium with the test organism was inoculated and incubated at 35-37°C up to 4 h, and examined at half hour intervals for a reduction reaction as shown by a change in color (compared with the positive control).

2.3. Plant Material and Preparation of Extracts

The air-dried leaves of *C. sophera* (25g) were extracted separately with petroleum ether, chloroform and ethanol at room temperature. The solutions were filtered and evaporated to dryness giving petroleum ether, chloroform and ethanol extracts (yields: 2.6, 1.91 and 2.2 g, respectively).

2.4. Antibacterial Assay

Antibacterial activity of the petroleum ether, chloroform and ethanol extracts leaves were individually tested against five enteric bacteria namely *Staphylococcus aureus*, *Pseudomonas aeruginosa*, *Enterococcus* species, *Klebsiella pneumonia* and *Escherichia coli*. Cultures of each bacterial strain were maintained on Luria-Broth (LB) medium (1.0% trypton, 0.5% yeast extract, 1.0% NaCl, 1.5% agar) at 4°C. Antibacterial test was then carried out by agar diffusion method [30], using 100 µl of standardized suspension of tested organisms (10^8 cfu ml^{-1}), spread on LB plates. The discs (6 mm in dia.) were impregnated with 10 µl of 30 mg ml^{-1} (300 µg $disc^{-1}$) plant extracts (extracts were dissolved in their respective solvents) and placed on the inoculated agar. Negative controls were prepared using the same solvents employed to dissolve the plant extracts. Streptomycin (30 µg disc-1) and tetracycline (30 µg $disc^{-1}$) were used as positive controls to determine the sensitivity of bacterial strain. The inoculated plates were incubated at 37°C for 24h. Antimicrobial activity was evaluated by measuring the zones of inhibition against the tested bacteria. Each assay was carried out in triplicate.

2.5. Minimum Inhibitory Concentration (MIC)

Minimum inhibitory concentration (MIC) of the crude (petroleum ether, chloroform and ethanol) extracts of *C. sophera* leaves were determined by two-fold serial dilution method [Chandrasekaran and Venkatesalu 2004]. Plant extracts were incorporated into 1 ml Luria-Broth medium to get a concentration of 1000 µg ml^{-1} and serially diluted to achieve 500, 250, 125, 62.5 and 31.25 µg ml^{-1}, respectively. 10 µl of standardized suspension of each tested organism (10^8 cfu ml^{-1}) was transferred to each tube. The control tubes contained only bacterial suspensions, were incubated at 37°C for 24 hours. The lowest concentration of the extracts, which did not show any growth of tested organisms after macroscopic evaluation, was determined as MIC.

3. RESULTS

3.1. Isolation and Characterization of Clinical Bacteria

The physical appearance of stool samples collected from both in patients of Kushtia General Hospital and pupils of various public primary schools in Kushtia, Bangladesh is shown in Table 1.

Gram-positive organisms were purple or blue-black with rod shaped, while Gram-negative organisms were red or pink with cocci shaped. Smaller (0.1-0.5 mm) and mucoid pink colony were catalase positive as they produced catalase enzyme; this enzyme converted hydrogen peroxide to water and oxygen gas was evolved. The evolution of gas caused bubbles to form, which indicated a positive test. Mucoid pink colony was also indicated a positive citrate test as they produced red color and indol test was negative which indicated the presence of *Klebsiella pneumoniae*. Smaller colony did not produce red color so it was *Staphylococcus aureus*. Smooth pink colony in indol test produced red surface layer, whereas in oxidase, catalase and citrate tests did not produce characteristic respective colors indicated *Escherichia coli*. Pale colored colony was changed into blue purple color within 30 seconds in oxidase test confirms the bacterium as *Pseudomonas aeruginosa*. Small dark red magenta colony under the litmus milk decolorization test produced white pale yellow color and unchanged in other tests, so it was *Enterococcus* sp. Finally, some clinical bacteria such as *Escherichia coli, Staphylococcus aureus, Enterococcus* sp., *Pseudomonas aeruginosa* and *Klebsiella pneumoniae* were isolated by color detection and colony size on MacConkey agar plate, as shown in Table 2.

Table 1. Physical appearance of stool samples collected from hospital patients and school pupils in Kushtia city of Bangladesh

Physical appearance	Number of samples from hospital patients	Number of samples from school pupils	Total number of collected samples
Samples with mucus	8 (27%)	6 (8.6%)	14
Samples with mixed mucus	12 (40%)	10 (14.3%)	22
Samples with visible blood	6 (20%)	4 (5.7%)	10
Samples without mucus and blood	4 (13%)	50 (71.4%)	54
Total	30	70	100

Table 2. List of bacteria identified by biochemical tests

Colony type	Gram stain	Cell Shape	Oxidase	Catalase	Indole	Citrate	Litmus milk decolori zation	Bacteria
Smaller (0.1-0.5 mm) colony	+	Cocci	-	+	ND	ND	ND	*S. aureus*
Small dark red magenta colony	+	Cocci	ND	-	ND	ND	+	*Enterococcus* sp.
Smooth pink	-	Rod	-	ND	+	-	ND	*E. coli*
Pale colored	-	Rod	+	-	-	+	ND	*P. aeruginosa*
Mucoid pink	-	Rod	-	+	-	+	ND	*K. pneumoniae*

ND= test not determined; + = 90% or more strains are positive; - = 90% or more strains are negative.

3.2. In Vitro Antibacterial Activity

In vitro antibacterial activity of the petroleum ether, chloroform and ethanol extracts of C. sophera leaves against the employed bacteria was qualitatively and quantitatively assessed by the presence or absence of inhibition zones. A total of five bacterial strains including two Gram-positive and three Gram-negative bacteria were tested. The chloroform and ethanol extracts revealed a great potential of antibacterial activity against all two Gram-positive and two Gram-negative bacteria at the concentration of 300 μg $disc^{-1}$ (Table 3).

The ethanol extract exhibited a potent inhibitory effect against *S. aureus*, *Enterococcus* sp., *Klebsiella pneumonia* and *E. coli* with their respective diameter zones of inhibition of 21.3 ± 1.1, 20.1 ± 0.9, 20.0 ± 0.7 and 18.2 ± 0.5 mm. However, *P. aeruginosa* was not inhibited. Chloroform extract also showed the strongest antibacterial effect against *Enterococcus* sp. (18.2 ± 0.5 mm), *S. aureus* (17.1 ± 0.7 mm), *Klebsiella pneumonia* (17.0 ± 1.1 mm) and *E. coli* (16.3 ± 0.7 mm).

Table 3. Antibacterial activity of *Cassia sopheria* L. leave extracts against enteric bacteria

Microorganism	Choloroform extract[a]	Petroleum ether extract[a]	Ethanol extract[a]	Antibiotic[b]	
				S	T
S. aureus	17.1 ± 0.7	12.1 ± 0.5	21.3 ± 1.1	16.2 ± 0.9	23.1 ± 1.5
Enterococcus sp.	18.2 ± 0.5	14.2 ± 1.1	20.1 ± 0.9	18.3 ± 0.7	18.1 ± 0.6
P. aeruginosa	nd	nd	nd	19.2 ± 1.1	17.3 ± 0.5
E. coli	16.3 ± 0.7	nd	18.2 ± 0.5	17.1 ± 0.5	20.2 ± 0.5
K. pneumoniae	17.0 ± 1.1	nd	20.0 ± 0.7	18 ± 0.4	21.1 ± 1.1

Values are represented as mean ± S.D. of triplicate experiments. nd: no detection.

[a] Diameter of inhibition zone including diameter of disc 6 mm (tested at a volume of 300 μg $disc^{-1}$).

[b] S: streptomycin (30 μg $disc^{-1}$); T: tetracycline (30 μg $disc^{-1}$).

Table 4. MIC values of *Cassia sopheria* L. leave extracts against enteric bacteria

Microorganism	Minimum inhibitory concentration (MIC)[a]		
	Choloroform extract	Petroleum ether extract	Ethanol extract
S. aureus	125	500	62.5
Enterococcus sp.	125	500	62.5
P. aeruginosa	nd	nd	nd
E. coli	250	nd	125
K. pneumoniae	125	nd	62.5

[a] Minimum inhibitory concentration (values in μg ml^{-1}). nd: no detection.

On the other hand, petroleum ether displayed a moderate inhibitory effect against two Gram-negative bacteria such as *S. aureus* and *Enterococcus* sp. with inhibition zones of in the range between 12.1 ± 0.5~14.2 ± 1.1 mm. The activity of the extracts was compared with two standard drugs such as streptomycin and tetracycline. In this study, in all cases, ethanol extract exhibited higher antibacterial activity compared to streptomycin, while tetracycline showed higher activity in some other cases than the ethanol extract and other solvent extracts. The blind control did not inhibit the growth of the bacteria tested. No inhibitory effect was observed against *P. aeruginosa* in all cases.

3.4. Minimum Inhibitory Concentration (MIC)

As shown in Table 4, the ethanol extract were found more susceptible to *Enterococcus* sp., *S. aureus*, *Klebsiella pneumonia* and *E. coli* with their MIC values ranging from 62.5~125 μg ml^{-1}. Strong inhibition of chloroform extract against *Enterococcus* sp., *S. aureus*, *Klebsiella pneumonia* and *E. coli* was also observed (MIC values of 125~250 μg ml^{-1}).

On the other hand, petroleum ether extract showed inhibitory activity against *Enterococcus* sp. and *S. aureus* (MIC value 500 μg ml^{-1} for each). In this study, the ethanol extract showed the highest antibacterial activity against the microorganisms tested, with the lowest MIC values.

DISCUSSION AND CONCLUSION

The increasing social and economic implications caused by pathogenic bacteria means there is a constant striving to develop new antibacterial agents. Due to the identified and potential toxicity of chemical antibiotics, there has been an increase demand for safe and effective antimicrobials from natural sources. The demand for more natural antimicrobials has driven scientists to investigate the effectiveness of inhibitory compounds such as extracts from plants [Nasar-Abbas and Halkman, 2004, Sohn, et al 2004]. Various publications have documented the antimicrobial activities of plant extracts [Ahmad and Aqil, 2007, Hwang, et al 2004, Mathabe, et al 2006]. Thus plant extracts are promising natural antimicrobial agents with potential applications in pharmaceutical industry for controlling the pathogenic bacteria.

In this present work, we found that the extracts of *C. sophera* significantly inhibited the growth of some enteric or pathogenic bacteria. The results of this study showed that chloroform and ethanol extracts of leaves had a great potential antibacterial activity against *S. aureus*, *Enterococcus* sp., *Klebsiella pneumonia* and *E. coli*. Petroleum ether extract also displayed moderate antibacterial activity against two Gram-positive bacteria such as *Enterococcus* sp. and *S. aureus*. However, *P. aeruginosa* was not inhibited.

When comparing the data obtained in different studies, most publications provide generalization about whether or not a plant extract possesses activity against Gram-positive and Gram-negative bacteria. However, not all provide details about the extent or the spectrum of this activity. Most antibacterial medicinal plants attack Gram- positive strains, while few are active against Gram-negative bacteria [Herrera, et al 1996, Scrinivasan, et al 2001]. In this study, we found that the chloroform and ethanol extracts were active against two Gram-positive and two Gram-negative strains out of five bacteria tested. However, a Gram-negative bacterium namely *P. aeruginosa* was not inhibited. In comparison with commercial antibiotics, the significant antibacterial activity was observed from most of the extracts. In maximum cases, the chloroform and ethanol extracts exhibited similar or higher antibacterial activity than standard drug streptomycin, whereas, tetracycline displayed higher activity in most of the cases than the plant extracts. Some pharmacological activities of ethanol extract of seed of *C. sophera* var. *purpurea* Roxb in rats has been reported earlier [Ahmad-Billal,et al 2005].

Thus, *C. sophera* could become an alternative to synthetic bactericides for using in pharmaceutical industries to control pathogenic bacteria. However, if plant extracts are to be used for medicinal purposes, issues of safety and toxicity will always need to be addressed.

REFERENCES

Agharkar, S. P. 1991. Medicinal Plants of Bombay Presidency, p. 58. Scientific Publishers, Jodhpur, India.

Ahmad, I. and Aqil, F. 2007. In vitro efficacy of bioactive extracts of 15 medicinal plants against ESβL-producing multidrug-resistant enteric bacteria. *Microbiological Research* 162: 264–275.

Ahmad-Billal, Naeem, A., Khan, A., Ghufran and Inamuddin. 2005. Pharmalogical Investigation of *Cassia sophera,* Linn. var. purpurea Roxb. *Medical Journal of Islamic World Academy of Sciences* 15 (3): 105–109.

Awan, M. H. 1984. Kitab-ul-Mufridat Almaruf be Khawas-ul-Advia, p. 372. Sheikh Ghulam Ali and Sons Ltd Publishers, Lahore, Pakistan.

Bhan, M.K. 2000. Current and future management of childhood diarrhoea. *International Journal of Antimicrobial Agents* 14: 71–73.

Chandrasekaran, M. and Venkatesalu, V. 2004 Antibacterial and antifungal activity of *Syzygium jambolanum* seeds. *Journal of Ethnopharmacology* 91: 105–108.

Chopra, R. N., Nayar, S. L. and Chopra, I. C. 1956. Glossary of Indian Medicinal Plants, p. 55. CSIR, New Delhi, India.

Conner, D. E. 1993. Naturally occurring compounds. pp. 441–468. In P. M. Davidson and A. L. Branen (eds.), Antimicrobials in foods. New York: Marcel Dekker.

Cowan, M. M. 1999. Plant Products as Antimicrobial Agents. *Clinical Microbiology Reviews* 12(4): 564-582.

Cox, P.A. and Balick, M.J. 1994. The ethnobotanical approach to drug discovery. *Scientific American* 270: 82–87.

Drury, H. 1873. Useful Plants of India, pp. 122-123. William H Allen and Co London.

Dutt, U. C. 1995. The Materia Medica of the Hindus, pp. 156-157. Mittal Publications, New Delhi.

Dymock, W., Warden, C.H. and Hooper, D. 1990. Pharmacographica India, pp. 520-523. Vol II. Trubner and Co, Ltd, London.

Herrera, R.M., Perez, M., Martin-Herrera, D.A., Lopez-Garcia, R. and Rabanal, R.M. 1996. Antimicrobial activity of extracts from plants endemics to the Canary Islands. *Phytotheraphy Research* 10: 364–366.

Hwang, J.K., Shim, J.S. and Chung, J.Y. 2004. Anticariogenic activity of some tropical medicinal plants against *Streptococcus mutans*. *Fitoterapia* 75: 596–598.

Katouli, M., Jaaffari, A., Farhoundi, A. and Ketabi, G. R. 1990. Aetiological studies of diarrhoea diseases in infants and young children in Iran. *J. Trop. Med. Hyg.* 9(3): 22 – 27.

Kirtikar, K. R. and Basu, B. D. 2000. Indian Medicinal Plants, pp. 1207-1210. Vol. IV. Sri Satguru Publications, Delhi.

Kotzé, M. and Eloff, J.N. 2002. Extraction of antibacterial compound from *Combretum microphyllum* (Combretaceae). *South African Journal of Botany* 68: 62–67.

Lubhaya, R. 1975. Goswami Bayanul Advia, Goswami Pharmacy, p.147. Vol. II. Qasim Jan Street, Delhi.

Mathabe, M.C., Nikolova, R.V., Lall, N. and Nyazema, N.Z. 2006. Antibacterial activities of medicinal plants used for the treatment of diarrhoea in Limpopo Province, South Africa. *Journal of Ethnopharmacology* 105: 286–293.

Nadkarni, K.M. 1982. Indian Materia Medica, pp. 290-291. Vol. I. Popular Parkashan Pvt. Ltd., Bombay.

Nasar-Abbas, S. M., and Halkman, A. K. 2004. Antimicrobial effect of water extract of sumac (*Rhus coriaria* L.) on the growth of some food borne bacteria including pathogens. *International Journal of Food Microbiology* 97: 63–69.

Njoku-Obi and Anozie, S.O. 1984. Outbreak of diarrhoea caused by enteropathogenic *E. coli* in a new born unit. *Nig. J. Microbiol.* 4(1): 15 –19.

Balandrin, M.F. (eds.), Human Medicinal Agents from Plants, ACS Symposium Series 534. American Chemical Society, Washington.

Rhodes, M.C.1996. Physiologically-active compounds in plant foods: an overview. *Proceedings of the Nutrition Society* 55: 371–384.

Scrinivasan, D., Nathan, S., Suresh, T., and Perumalsamy, O., 2001. Antimicrobial activity of certain Indian medicinal plants used in folk loric medicine. *Journal of Ethnopharmacology* 74: 217–220.

Sohn, H.Y., Son, K.H., Kwon, C.S., and Kang, S.S. 2004. Antimicrobial and cytotoxic activity of 18 prenylated flavonoids isolated from medicinal plants: *Morualba* L., *Morumongolica* Schneider, *Broussentia papyrifera* (L.) Vent, *Sophora flavescens* Ait and *Echinosophora koreensis* Nakai. *Phytomedicine* 11: 666–672.

World Health Organisation. 1996. Water and Sanitation: Fact sheet. Vol. 112. Geneva, World Health Organisation.

In: Characterization and Development of Novel Materials
Editor: Rafiq Islam
ISBN: 978-1-61324-149-3

Chapter 7

THE FORMULATION OF COMPREHENSIVE MASS AND ENERGY BALANCE EQUATION TOWARDS DEVELOPING SUSTAINABLE MODEL WITH THE SCIENCE OF INTANGIBLES

M. Moniruzzaman Khan**, Chefi Ketata and M. Rafiqul Islam
Department of Civil and Resource Engineering, Dalhousie University
Halifax, Canada

ABSTRACT

Today's civilization is based on the technologies that are tangible in nature. However, tangibles relate to short-term and very limited in space whereas the intangibles relate to either long-term or to other elements of the current time frame. Therefore, a focus on tangible will continue to obscure long-term consequences. The long-term consequences will not be uncovered until intangible properties are properly analyzed and included. This study has emphasized on intangibles due to its inherent importance and showed how tangibles should link with intangibles. This has opened up the understanding of the relationship between intangible to tangible scales; from microscopic to macroscopic properties. This study also develops a guideline for sustainability using mass and energy balance which is an extended work of sustainability development by Khan and Islam, 2007a (Khan, M.I and Islam, M.R., (2007) True Sustainability Technological Development and Natural Resources Management, Nova Science Publishers, NY, USA, 381p.). The zero-waste concept is inherently related to true sustainability. In this study, a detailed analysis of different features of sustainability is presented in order to understand the importance of using the concept of sustainability in every technology development model. It was identified that for a process to be sustainable, it must have the source as well as the pathway be natural. Observation and understanding of nature, mass and energy balance and the relationship, activities etc. were the research tools to explain the importance of intangibles in human life and technology development. The current study takes a significant step forward developing the link between 'hard science' and the science of intangibles.

* Corresponding author: E-mail: mmkhan@dal.ca.

Keywords: sustainability; nature; intangibles; characteristic time; mass balance.

1. INTRODUCTION

The term 'sustainable' is a fast-growing concept of today's technology development. Commonly, the use of this term infers that the process is beneficial to human beings, both in short-term and long-term. The concept 'zero waste living' has been generated from the true sustainable criterion which is derived from the undisturbed activities of nature. Scientifically, this means that true long-term considerations of humans should include the entire ecosystem. Some have called this inclusion 'humanization of the environment' and put this phenomenon as a pre-condition to true sustainability (Zatzman and Islam, 2007b). The inclusion of the entire ecosystem is only meaningful when the natural pathway for every component of the technology is followed. Only such design can assure both short-term (tangible) and long-term (intangible) benefits.

Recently, Chhetri (2006) has established that by taking long-term (intangible) approach the outcome is reversed from the one that emerges from short-term (tangible) approach. He made this observation in relation to energy efficiency of various energy sources. By focusing on just heating value, one comes up with a ranking that diverges into what is observed as the global warming phenomenon. On the other hand, if long-term approach was taken, none of the previously perpetrated technologies would be considered 'efficient' and would long been replaced with truly efficient (global efficiency-wise) technologies, avoiding the currently faced energy crisis. Zatzman and Islam (2007a) equated real with natural and further argued that sustainability of an action is similar to sustainability of a process. An action to be sustainable (or real), it must have its origin real to begin with. If the origin of any action is the intention, they argued, intention must play a role in defining success of an action, and subsequently, the process that is developed by these actions.

It has long been accepted that Nature is self-sufficient and complete, rendering it as the true teacher of how to develop sustainable technologies. From the standpoint of human intention, this self-sufficiency and completeness is actually a standard for declaring Nature perfect. "Perfect" here, however, does not mean that Nature is in one fixed unchanging state. On the contrary, it is the capacity of Nature to evolve and sustain that makes it such an excellent teacher. This perfection makes it possible and necessary for Humanity to learn from Nature not to fix Nature but to improve its own condition and prospects within Nature, in all periods and for any timescale. The significance of such emulating of Nature is subtle but crucial: it is that technological or other development undertaken within the natural environment only for some limited short term must necessarily, sooner or later, end up violating something fundamental or characteristic within Nature. Understanding the effect of intangibles and the relations of intangible to tangible is important for reaching appropriate decisions affecting the welfare of the society and the nature as well. A number of aspects of natural phenomena have been discussed here to find out the relationship between intangible and tangible. The target of this study is to provide strong basis to the sustainability model. The mass and energy balance equation has provided and explained to support the influence of intangibles and the role of intention.

2. Development of a Sustainable Model

The sustainability model developed by Khan and Islam (2007a) provides the basis of the direction of the sustainable technology. According to this model, a process is sustainable if and only if it travels a path that is beneficial for an infinite span of time. Otherwise the process must diverge in a direction that is not beneficial in the long run. Pro-nature technology is the long-term solution; anti-nature solutions are all that will come from schemas that comprehend, or analyze, or plan to handle change on the basis of any approach in which time-changes, or Δt, are examined only as they approach 0 (zero) that have been designed or selected as being good for time t ='right now' (equivalent to the idea of $\Delta t \rightarrow 0$). Of course, in nature, time "stops" nowhere, there is no such thing as steady state, and hence, regardless of the self-evident tangibility of the technologies themselves, the "reality" in which they are supposed to function usefully is itself non-existent – what can be described as "aphenomenal" – and cannot be placed on the graph (Figure 1). "Good" technology can be developed – if and only if it travels a path that is beneficial for an infinite span of time. In Figure 1, this concept is incorporated in the notion of '*time tending to Infinity*', which (among other things) implies also that time-changes instead of approaching 0 (zero) could instead approach Infinity, *i.e.*, $\Delta t \rightarrow \infty$. In this study, it has been introduced the term 'perception' which was found important at the beginning of any process. Perception varies person to person. It is very subjective and there is no way to prove if a perception is true or wrong; its effect is immediate. Perception is completely one's personal opinion developed from one's experience without appropriate knowledge. That is why perception can not be used as the base of the model. However, if perception is used in the model, the model would look like as follows (Figure 2).

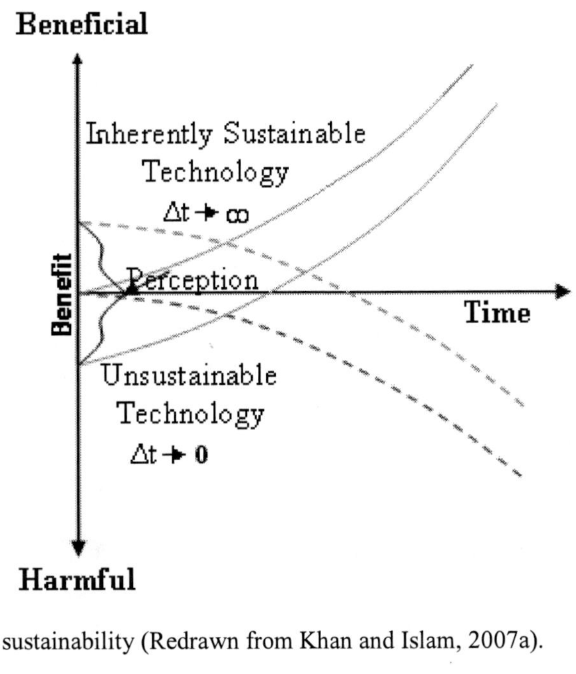

Figure 1. Direction of sustainability (Redrawn from Khan and Islam, 2007a).

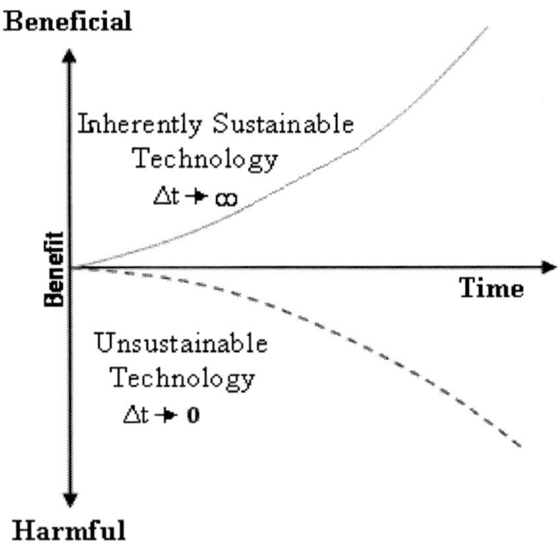

Figure 2. Direction of sustainability (Modified from Khan and Islam, 2007a).

2.1. Problem with the Current Model

Any current model of technology development focuses only to transient tangible benefit which falls into unsustainable technology at the end. For example, conventional engineering analysis today, in which electrical power generation is routine, common and widespread, would condemn as unsustainable the technology that converts the rotary motion of a wheel fixed in position over a flowing water course (river or stream) into useful work in the form of mechanical energy available immediately within a workplace constructed around the installation of this water wheel. However, every component of this arrangement can be renewed, replaced or maintained entirely from naturally available materials (wood for the wheel, keeping the water flow source undimmed or otherwise unobstructed), whereas vast tracts of nature have to be destroyed one way or another to sustain electrical power generation. Instead of being addressed keeping sustainability in mind, the limitations of the older technology's speed and scale of productivity were leap-frogged by diverging onto the anti-nature path. A century later the headlines would read: "Does Power Corrupt?" in an attempt to signal the devastating side effect of electrical energy (Mittelstaedt, 2007). The very fact that alternate current does not exist in nature should have alerted scientists that electrical energy is not sustainable. Instead, electrical energy has been made into norm and today's civilization is measured by how many buildings are light during the night (Figure 3). In this, even steam energy is marginalized unless turbines are run with nuclear energy.

A single analysis would have indicated decades ago that electrical heat or electrical light cannot be beneficial for the human being. However, this analysis would require researching beyond the immediate tangible and in the modern age there seems to be giant step backward toward tangibles. Energy technologies that survived the commercialization test ended up making things more focused on tangible (Khan and Islam, 2007b).

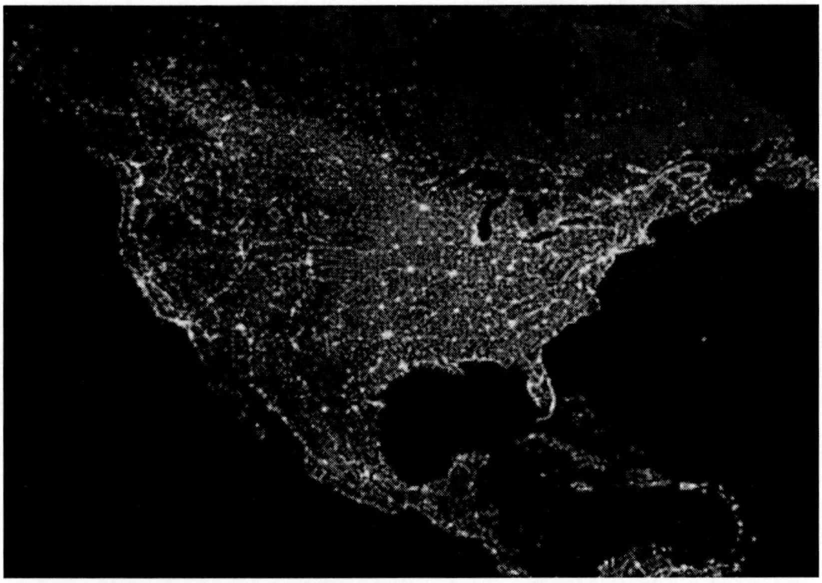

Figure 3. USA at night by NASA (soulincode, 2006).

With this mode, not only every technology developed is unsustainable but is also attached to a false promise. With time this promise became unraveled (Table 1).

If the products are analyzed with the implementation years and the time required to be exhausted, it is found that no non-natural products are sustainable (Figure 4). The effect of some products continues even after extinction of those products. For example, chemical such as DDT is still found in bird dropping (Guano) though DDT is no more practiced. These droppings stimulate the growth of mosses and plankton in the ponds, which feed various insects, which in turn support small birds, called snow buntings (Schmid, 2005).

Therefore, it is found that some pollutants never end due to their biological connections. Microoven, simulated wood, non-stick cooking jar, Freon, DDT, PVC etc were flourished during last several decades with lots of promises, however, those promises became falsehood with progressing time (Table 1). For instance, Freon changed its facade after 57 years, DDT after 30 years, PVC after 30 years; non-stick Teflon after 50 years; Enron after 15 years and VIOXX after 3 years; everything that is untrue will change its front after a finite time (DDT, 2006; PVC, 2006; Teflon, 2006; Enron, 2006 and VIOXX, 2006). Truth will not. In that sense, truth is the only steady state. However, none of the current technology with transient tangible model is steady and therefore, based on false promises (Zatzman and Islam, 2007a).

In this decade, Australia and some other countries have banned the use of incandescent light bulb which produces artificial light (Maclaren, 2007). Teflon is considered to be used as a non toxic, low frictional surface. However, its constituents are cancer causing. Teflon degrades at high temperature and that is why its uses in cookeries are alarming in the future. Even with the technological improvement, at least 70 percent of all new drugs introduced in the United States in the past 25 years come from nature despite the use of sophisticated techniques to design products in the laboratory (Steenhuysen, 2007). Their study indicates that a back-to-nature approach might yield better possibilities for companies looking for the next blockbuster drug.

Table 1. Analysis of "breakthrough" technologies (Modified from Khan and Islam, 2007a)

Product	False Promise (knowledge at t= 'right now')	Observation with Current Knowledge (closer to reality)
Microwave oven	Instant cooking (bursting with nutrition)	97% of the nutrients destroyed; produces dioxin from baby bottles
Fluorescent light (white light)	Simulates the sunlight and can eliminate 'cabin fever'	Used for torturing people, causes severe depression
Prozac (the wonder drug)/Zoloft	80% effective in reducing depression	Increases suicidal behavior
Anti-oxidants	Reduces aging symptoms	Gives lung cancer
Vioxx	Best drug for arthritis pain, no side effect	Increases the chance of heart failure
Coke	Refreshing, revitalizing	Dehydrates; used as a pesticide in India
Transfat	Should replace saturated fats, incl. high-fiber diets	Primary source of obesity and asthma
Simulated wood, plastic gloss	Improve the appearance of wood	Contains formaldehyde that causes Alzheimer
Wood processing	Weather resistant wood	Contain arsenic that causes lung and skin cancer
Cell phone	Empowers, keep connected	Gives brain cancer, decreases sperm count among men.
Chemical hair colors	Keeps young, gives appeal	Gives skin cancer
Chemical fertilizer	Increases crop yield, makes soil fertile	Harmful crop; soil damaged
Chocolate and 'refined' sweets	Increases human body volume, increasing appeal	Increases obesity epidemic and related diseases
Pesticides, MTBE	Improves performance	Damages the ecosystem
Desalination	Purifies water	Necessary minerals removed
Wood paint/varnish	Improves durability	Numerous toxic chemicals released
Leather technology	Won't wrinkle, more durable	Toxic chemicals
Freon, aerosol, etc.	Replaced ammonia that was 'corrosive'	Global harms immeasurable and should be discarded
Non-Stick (Teflon)	Replaced metal surface that is sticky	produces dioxins, especially at high temperature
DDT	Miracle powder, kills pathogens and harmful insects	Carcinogenic, inherently toxic
PVC	Healthy alternative to expensive wood materials	Carcinogenic, inherently toxic

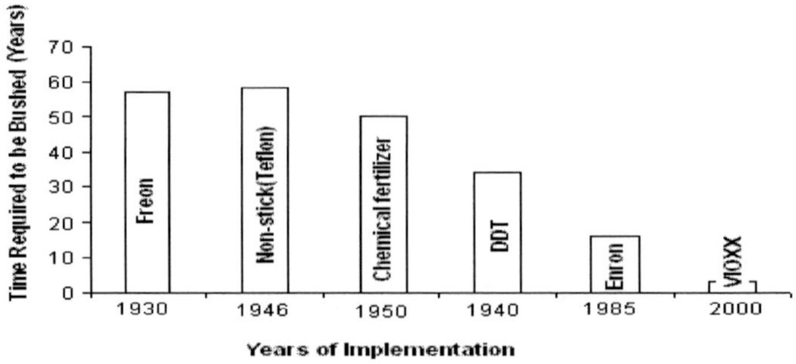

Figure 4. Bushed out time of false promised products.

2.2. Violation of Characteristic Time

Another problem of the current technology is that it violates the natural characteristic time. The characteristics time is similar to natural life cycle to any living being. However, characteristics time does not include any modification of life cycle time due to non-natural human intervention. For instance, the life span of an unconfined natural chicken can be up to 10 years, yet table fowls or broilers reach adult size and are slaughtered at six weeks of age (PAD, 2006). The characteristics time for broiler chicken has been violated due to human intervention. This study has emphasized on characteristic time because of its pro-nature definition. Anything found in the nature grown and obtained naturally is reached to its contents both in tangibles and intangibles. However, anything produced either by internal genetically intervention or external chemical fertilizer along with pesticide utilization can not be good for human consumption for both long term and short term benefit. The notion of this violation is only to produce more to obtain short-term tangible benefits trading off with other intangible benefits which are more important.

3. Observation of Nature: Importance of Intangibles

Nature is directly observed and recorded only in tangible aspects detectable with current technologies. Accordingly, much of what could only be taking place as a result of intangible but very active orderliness within nature is promoted – according to this exclusively and narrowly tangible standard – as being "disorder". The greatest confusion is created when this misapprehension is then labeled "chaotic," and its energy balance on this basis portrayed as headed towards "heat death", or "entropy", or the complete dissipation of any further possibility of extracting "useful work".

Reality is quite different. In nature, there is not a single entity that is linear, symmetric or homogeneous. In Globe, there isn't a single process that is steady or even periodic. Natural processes are chaotic, but not in the sense of being either arbitrary or inherently tending towards entropy. Rather, they are chaotic in the sense that what is essentially orderly and characteristic only unfolds with the passage of time within the cycle or frequency that is characteristic of the given process at some particular point. What the process looks like at that point is neither precisely predictable previous to that point nor precisely reconstructible or reproducible after that point: the path of such a process is defined as chaotic on the basis of its being a periodic, non-linear and non-arbitrary.

Nature is chaotic. However, the laws of motion developed by Newton can not explain the chaotic motion of Nature due to its assumptions which contradicts with the reality of Nature. The experimental validity of Newton's laws of motion is only limited to describe instantaneous macroscopic and tangible phenomena, however, microscopic and intangible phenomena are ignored. The classical dynamics, as represented by Newton's laws of motion, emphasizes fixed and unique initial conditions, stability, and equilibrium of a body in motion (Ketata et al., 2007a). However, the fundamental assumption of constant mass alone is adequate to conflict Newton's laws of motion. Ketata et al. (2007a) formulated the following relation to describe the body in continuous motion in one space:

$$m = \frac{F}{\left((6t+2)+\left(3t^2+2t+1\right)^2\right)ce^u} \qquad (1)$$

where

F is the force on the body;

$u = t^3 + t^2 + t + 1$;

and c is a constant.

The above relation demonstrates that the mass of a body in motion depends on time whether *F* varies over time or not. This is absolutely the contradiction of first law of motion. Similarly, the acceleration of a body in motion is not proportional to the force acting on the body because mass is not constant. Again, this is a contradiction of second law of motion.

Here it is found that time is the biggest issue which, in fact, dictates the correctness of Newton's laws of motion. Considering only instantaneous time ($\Delta t \rightarrow 0$), Newton's laws of motion will be experimentally valid with some error, however, considering the infinite time span ($\Delta t \rightarrow \infty$), the laws can not be applicable. That is why sustainable technologies which include short term to long term benefit can not be explained by Newton's laws. To overcome this difficulty, it is necessary to break out of '$\Delta t \rightarrow 0$', in order to include intangibles, which is the essence of pro-nature technology development.

In terms of the well-known laws of conservation of mass (*m*), energy (*E*) and momentum (*p*), the overall balance, *B*, within Nature may be defined as some function of all of them:

$$B = f(m, E, p) \qquad (2)$$

The perfection without stasis that is Nature means that everything that remains in balance within it is constantly improving with time. That is:

$$\frac{dB}{dt} > 0. \qquad (3)$$

If the proposed process has all concerned elements such that each element is following this pathway, none of the remaining elements of the mass balance discussed later will present any difficulty. Because the final product is being considered as time extends to infinity, the positive ("> 0") direction is assured.

Pro-nature technology, which is non-linear, increases its orderliness on a path that converges at infinity after providing maximum benefits over the intervening time. This is achievable only to the extent that such technologies employ processes as they operate within nature, or use materials whose internal chemistry has been refined entirely within the natural environment and whose subsequent processing has added nothing else from nature in any manner other than its characteristic form. Any and every other technology is anti-nature. The worst among them are self-consciously linear, "increasing" order artificially by means of successive superpositions that supposedly take side-effects and negative consequences into account as they are detected. This enables the delivery of maximum power, or efficiency, etc. for some extremely short term, without regard to coherence or overall sustainability and at the cost of manifold detrimental consequences carrying on long after the "great advances" of the original anti-nature technology have dissipated. Further disinformation lies with declaring the resulting product 'affordable', 'inexpensive', 'necessary' and other self serving and utterly false attributes while increasing only very short terms costs. Any product that is anti-nature would turn out to be prohibitively costly if long-term costs are included. The case in point is the tobacco technology. In Nova Scotia alone, 1300 patients die each year of cancer emerging directly from smoking (Islam, 2003). These deaths cost us 60 billion dollars in body part alone. How expensive cigarette should be? The fact is, if intangibles are included in any economic analysis, a picture very different from what is conventionally portrayed will emerge (Zatzman and Islam, 2007b).

Any linearized model can be limited or unlimited, depending on the characteristics of the process (Figure 5). The "limited linearized model" has two important characteristics: more tangible features than intangible, and a finite, limited amount of disorder or imbalance. Because only linearized models are man-made, nature has time to react to the disorder created by this limited model, and it may, therefore, be surmised that such models are unlikely to cause damage that is irreparable.

The unlimited linearized model is characterized by long-term effects little understood but far more damaging, with more intangible features than tangible and an unlimited degree of disorder, or imbalance. Contemporary policy-making processes help conceal a great deal of actual or potential imbalance from immediate view or detection – a classic problem with introducing new pharmaceuticals, for example. Since a drug has to pass the test of not showing allergic reactions, many such drugs make it into the market after being "tweaked" to delay the onset of what are euphemistically called "contra-indications."

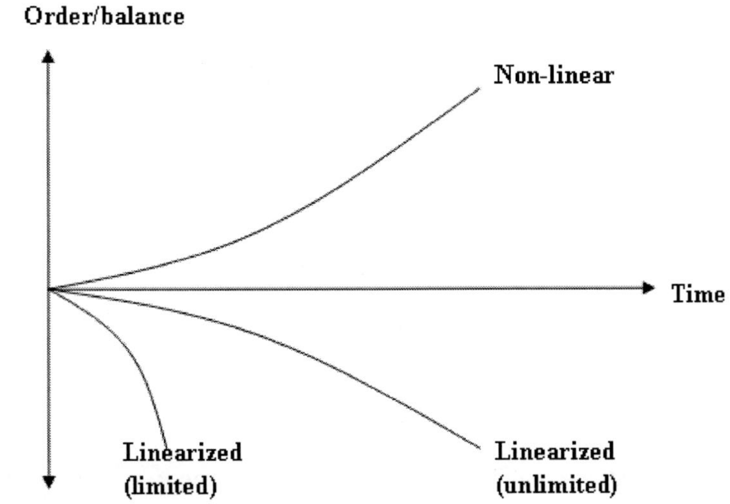

Figure 5. Pathway of nature and anti-nature (Modified from Khan and Islam, 2007a).

An elaborate and tremendously expensive process of clinical trials is unfolded to mask such "tweaking", mobilizing the most heavily invested shareholders of these giant companies to resist anything that would delay the opportunity to recoup their investment in the marketplace. The growing incidence of suicide among consumers of Prozac® and other SSRI-type anti-depressant drugs, and of heart-disease "complications" among consumers of "Cox-2" type drugs for relief from chronic pain, are evidence of the consequences of the unlimited linearized model and of how much more difficult any prevention of such consequences is (Miralai, 2006). In forms of concentrations, unlimited pertains to intangible.

Here is another "homier" example of how the unlimited linearized model delays the appearance of symptoms: if food is left outside, in 2-3 days, it will cause food poisoning provoking diarrhea. However, if the food is placed in artificial refrigeration, the food will retain some appearance of 'freshness' even after several weeks – although its quality will be much worse than the 'rotten' food that was left outside. Another more exotic but non-industrial example can be seen in the reaction to snake venom. The initial reaction is immediate. If the victim survives, there is no long-term negative consequence. Used as a natural source or input to a naturally-based process, snake venom itself possesses numerous long-term benefits and is known for its anti-depressed nature.

Repositioning cost-benefit analysis away from such short-term considerations as the cheapness of synthesized substitutes, etc to the more fundamental tangible/intangible criterion of long-term costs and benefits, the following summary emerges: tangible losses are very limited, but intangible losses are not.

4. Analogy of Physical Phenomena

Mathematicians continue to struggle with the two entities '0' and '∞', whose full meanings and consequences continue to mystify (Ketata et al., 2006a; 2006b). However, these two entities are most important when intangible issues are counted, as the following simple analogy from well-known physical phenomena (Figure 6) can demonstrate.

As "size", *i.e.*, space occupied (surface area or volume) per unit mass, goes down; the quantity of such forms of matter goes up. This quantity approaches infinity as space occupied per unit mass heads towards zero; however, according to the Law of Conservation of Mass and Energy, mass can neither be created nor destroyed and it only can transform from one form to another form. This immanent contradiction was resolved in the early 20th century, as it was proven that as mass decreased; its quantity could increase as particles of mass were converted into quanta of energy. Today, up to 60 particles have been identified and the search for more continues (He, 2005). In the same vein, as the size of the particle decreases the particle-particle interaction energy changes dramatically (Ball, 2007). For this analysis to be continuous, as stipulated by the nature of Nature, the curve in Figure 6 should extend up to infinity.

Infinity means that a quantity too large to count exactly, but that enjoys practical existence. Conventionally, zero on the other hand denotes non-existence, posing another paradox that is nonetheless removable… when the intangible aspect is considered. Some thing that is infinite in number is present everywhere but has no size.

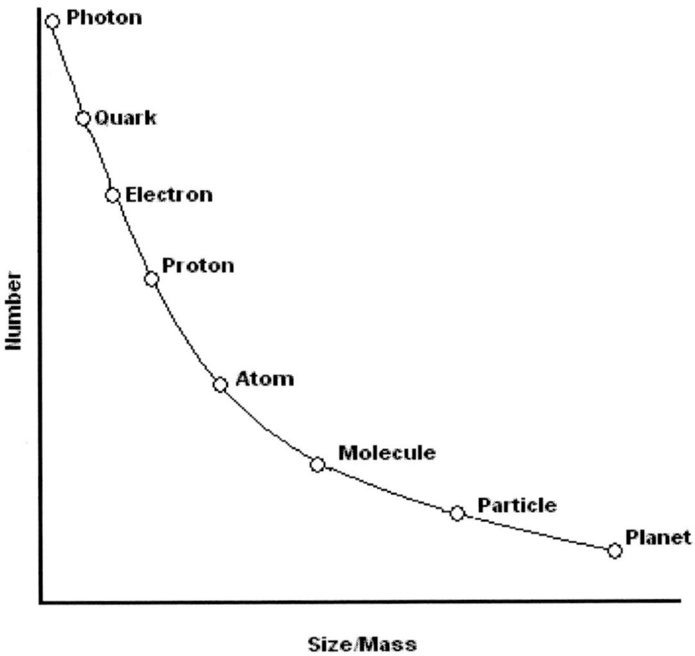

Figure 6. Relation of Size/Mass to number.

As Figure 6 shows, mass turns into the energy at the end – losing “size”, a transition of the tangible into the intangible. This also signifies that the number of intangibles is much more than that of tangibles. We can measure the tangible properties but it is difficult to measure the intangible. Yet, inability to measure the intangible hardly demonstrates non-existence. Happiness, sorrow, etc are all clearly intangible, and – however tangible their causes –possess no tangible properties whatsoever in themselves. As Figure 5 suggests, the scale of the intangible is potentially far more consequential than that of the tangible.

5. Intangible Cause to Tangible Consequence

Short-term intangible effects are difficult to understand, but consideration of the treatment procedures employed by homeopaths may serve to illustrate. The most characteristic principle of homeopathy is that the potency of a remedy can be enhanced by dilution, an inconsistency with the known laws of chemistry (Homeopathy, 2006). In some case, the dilution is so high that it is extremely unlikely that one molecule of the original solution would be present in that dilution. As there is no detectable mechanism to this, the effect of the molecule cannot always be understood and that is why the homeopathy still remains controversial to the modern science of tangible. However, the trace ingredient of dilution is not always ignorable. Recently, Rey (2003) studied the thermoluminescence of ultra-high dilution of lithium chloride and sodium chloride and found the emitted light specific of the original salts dissolved initially. The dilution was beyond Avogadro’s number ($\sim 6.0 \times 10^{23}$ atoms per mole) – but its effect was visible. In other words: when concentration of a substance descends to below detection level, it cannot be ignored, as its effects remain present. This is where greater care needs to be taken in addressing the harmful potential of chemicals in low concentrations. Lowering the concentration cannot escape the difficulty – a significant consideration when it comes to managing toxicity. Relying on low concentration as any guarantee of safety defeats the purpose when the detection threshold used to regulate as to what is “safe” is itself higher than the lowest concentrations at which these toxins may be occurring or accumulating in the environment. Although the science that will identify the accumulation of effects from toxic concentrations before they reach the threshold of regulatory detection remains to be established, the point is already clear: tangible effects may proceed from causes that can remain intangible for some unknown period of time.

Mobile phones are considered to be one of the biggest inventions of modern life for communication. So far, the alert of using mobile phone was limited only to the human brain damage for non-natural electro magnetic frequency. An official Finnish study found that people who used the phones for more than 10 years were 40 per cent more likely to get a brain tumor on the same side as they held the handset (Lean and Shawcross, 2007). However, recently it has been observed that mobile frequency also cause serious problem to the other living being of nature, which are very important for use to balance the ecological system. Recently, an abrupt disappearance of the bees that pollinate crops has been noticed especially in USA as well as some other countries of Europe (Lean and Shawcross, 2007). The plausible explanation of this disappearance is that radiation from mobile phones interferes with bees' navigation systems, preventing the famously home loving species from finding their way back to their hives. Most of the world's crops depend on pollination by bees. That is why a massive food shortage has been anticipated due to the extinction of these bees due to radiation given off by the mobile phones. Albert Einstein once said that if the bees disappeared, "man would have only four years of life left" (Lean and Shawcross, 2007). This is how a non-natural hi-tech instrument poses tangible effect in the long run due to its intangible causes.

6. Removable Discontinuities: Phases and Renewability of Materials

By introducing time-spans of examination unrelated to anything characteristic of the phenomenon itself being observed in nature, discontinuities appear. These are entirely removable, but they appear to the observer as finite limits of the phenomenon itself, and as a result, the possibility that these discontinuities are removable is not even considered. This is particularly problematic when it comes to the matter of phase transitions of matter and the renewability or non-renewability of energy.

The transition between solid, liquid and gas in reality is continuous – but the analytical tools formulated in classical physics are anything-but: each P-V-T model applies to only one phase and one composition, and there is no single P-V-T model applicable to all phases (Cismondi and Mollerup, 2005). Is this an accident? Microscopic and intangible features of phase-transitions have not been taken into account, and as a result of thus limiting the field of analysis to macroscopic, entirely tangible features, modeling becomes limited to one phase and one composition at a time.

When it comes to energy, everyone has learned that it comes in two forms: renewable and nonrenewable. If a natural process is being employed, however, everything must be "renewable" by definition – in the sense that, according to the Law of Conservation of Energy, energy can be neither created nor destroyed. Only the selection of the time-frame misleads the observer into confounding what is accessible in that finite span with the idea that energy is therefore running out. The dead plant material that becomes petroleum and gas trapped underground in a reservoir is being added to continually, but the rate at which it is extracted has become set according to an intention that has nothing to do with what the optimal timeframe in which the organic source material could be renewed. Thus, "non-renewability" is not any kind of absolute fact of nature. On the contrary: it amounts to a declaration that the pathway on which the natural source has been harnessed is anti-Nature.

7. Rebalancing Mass and Energy

Mass and energy balance inspected in depth disclose intention as the most important parameter, as the sole feature that renders the individual accountable to, and within, nature. This is rife with serious consequences for the black-box approach of conventional engineering, because a key assumption of the black-box approach stands in stark and howling contradiction to one of the key corollaries of that most fundamental principle of all, the Law of Conservation of Matter.

Conventionally, the mass balance equation is represented as "mass-in equals mass-out" (Figure 7). In fact, however, this is only possible if there is no leak anywhere and no mass can flow into the system from any other point – thereby rendering the entire analysis a function of tangible measurable quantities, a "science" of tangibles-only, like South Africa without mineworkers and field-hands before 1990.

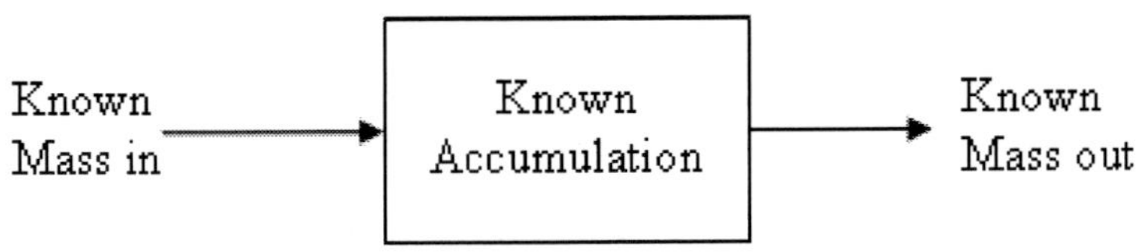

Figure 7. Conventional Mass Balance equation incorporating only tangibles.

The mass conservation theory indicates that the total mass is constant. It can be expressed as follows:

$$\Sigma_0^{\infty} m_i = \text{Constant} \quad (4)$$

where m = mass and i is the number from 0 to ∞.

In true sense, this mass balance encompasses mass from very macroscopic to microscopic and detectable to undetectable, *i.e.*, from tangible to intangible. Therefore, the true statement should be as illustrated in Figure 8:

"***Known*** mass-in" + "***Unknown*** mass-in" = "***Known*** mass-out +"***Unknown*** mass-out" +"***Known*** accumulation" + "***Unknown*** accumulation" (5)

The unknown masses and accumulations are neglected, which means they are considered to be equal to zero.

Every object has two masses:

(1) Tangible mass.
(2) Intangible mass usually neglected.

Then, Equation (5) becomes:

$$\sum_{i=0}^{\infty} m_{in,i} + \sum_{i=0}^{\infty} 0_{m_{in,i}} = \sum_{i=0}^{\infty} m_{out,i} + \sum_{i=0}^{\infty} 0_{m_{out,i}} + \sum_{i=0}^{\infty} m_{acc,i} + \sum_{i=0}^{\infty} 0_{m_{acc,i}} \quad (6)$$

The unknowns can be considered intangible, yet essential to include in the analysis as they incorporate long-term and other elements of the current timeframe.

In nature, the deepening and broadening of order is continually observed, with many pathways, circuits and parts of networks being partly or even completely repeated and the overall balance being further enhanced. Does this actually happen as arbitrarily as conventionally assumed? A little thought suggests this must take place principally as a result and-or in response to human activities and the response of the environment to these activities and their consequences.

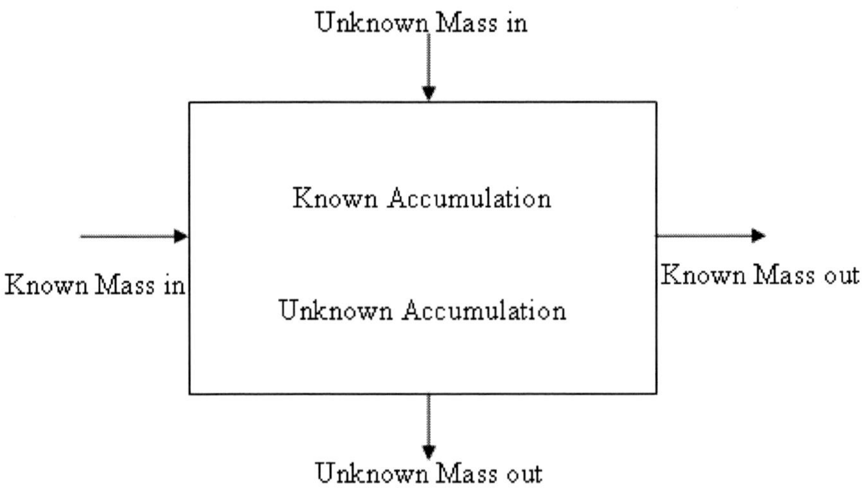

Figure 8. Mass-balance equation incorporating tangibles and intangibles.

Nature itself has long established its immediate and unbreachable dominion over every activity and process of everything else in its environment, and there is no other species that can drive nature into such modes of response. In the absence of the human presence, nature would not be provoked into having to increase its order and balance, and everything would function in "zero net waste" mode.

An important corollary of the Law of Conservation of Mass, that mass can be neither created nor destroyed, is that there is no mass that can be considered in isolation from the rest of the universe. Yet, the black-box model clearly requires just such an impossibility. Since, however, human ingenuity can select the time frame in which such a falsified "reality" will be exactly what the observer perceives, the model of the black box can be substituted for reality and the messy business of having to take intangibles into account is foreclosed once and for all.

8. ENERGY: EXISTING MODEL

There are a number of theories developed in the past centuries to define energy and its characteristics. However, none of the theories is enough to describe energy properly. All of the theories are based on much idealized assumptions which never exist practically. Consequently, the existing model of energy and its relation to others can not be accepted confidently. For instance, the second law of thermodynamics depends on Carnot cycle in the classical thermodynamics, where none of the assumptions of Carnot's cycle exists in reality. Definition of ideal gas, reversible process and adiabatic process used in describing the Carnot's cycle are imaginary. In 1905, Einstein came up with his famous equation, $E=mc^2$ which states an equivalence between energy (*E*) and relativistic mass (*m*), in direct proportion to the square of the speed of light in a vacuum (c^2). However, the assumptions of constant mass and the concept of vacuum do not exist in reality. Moreover, this theory was developed on the basis of planks constant which was derived from black body radiation. Perfectly black body is not even existed in reality. So it is found that, the development of every theory is dependent of a series of assumptions which do not exist in reality.

8.1. Supplements of Mass Balance Equation

For whatever else remains unaccounted, the mass balance equation, which in its conventional form necessarily falls short of explaining the functionality of nature coherently as a closed system, is supplemented by the energy balance equation.

For any time, the energy balance equation can be written as:

$$\int_0^\infty \Sigma a_i = \text{Constant} \quad , i \text{ going from 1 to infinity} \tag{7}$$

where *a* is the activity equivalent to potential energy.

In the above equation, only *potential* energy is taken into account. Total potential energy, however, must include all forms of activity, and here once again a large number of intangible forms of activity, e.g., the activity of molecular and smaller forms of matter, cannot be "seen" and accounted in this energy balance. The presence of human activity introduces the possibility of other potentials that continually upset the energy balance in nature. There is overall balance but some energy forms, *e.g.*, electricity (either from combustion or nuclear sources), which would not exist as a source of useful work except for human intervention, continually threaten to push this into a state of imbalance.

In the definition of activity, both time and space are included. The long term is defined by time being taken to infinity. The "zero waste" condition is represented by space going to infinity. There is an intention behind each action and each action is playing important role in creating overall mass and energy balance.

The role of intention is not to create a basis for prosecution or enforcement of certain regulations. It is rather to provide the individual with a guideline. If the product or the process is not making things better with time, it is fighting nature – a fight that cannot be won and is not sustainable. Intention is a quick test that will eliminate the rigorous process of testing feasibility, long-term impact, etc. Only with 'good' intention things can improve with time. After that, other calculations can be made to see how fast the improvement will take place.

In clarifying the intangibility of an action or a process, with reference to the curve of Figure 1, the equation has some constant which is actually an infinite series:

$$a = \Sigma_0^\infty a_i = a_0 + a_1 + a_2 + a_3 + \ldots\ldots.. \quad (8)$$

If each term of Eq. 8 converges, it will have a positive sign, indicating intangibility; hence the effect of each term thus becomes important for measuring the intangibility overall. On this path, it should also become possible to analyze the effect of any one action and its implications for sustainability overall as well.

In can be inferred that man-made activities are not enough to change the overall course of nature. Failure up to now, however, to include an accounting for the intangible sources of mass and energy has brought about a state of affairs in which, depending on the intention attached to such intervention, the mass-energy balance can either be restored and maintained over the long-term, or increasingly threatened and compromised in the short term. In the authors' view, it would be far better to develop the habit of investigating Nature and the prospects and possibilities it offers Humanity's present and future by considering time t at all scales, going to infinity, and give up once and for all the habit of resorting to time scales that appear to serve some immediate ulterior interest in the short term but that in fact have nothing to do with the natural phenomena of interest and must therefore lead to something that will be anti-Nature in the long-term and the short-term.

The main obstacle to discussing and positioning the matter of human intentions within the overall approach to the Laws of Conservation of Mass, Energy and Momentum stems from notions of the so-called "heat death" of the universe predicted in the 19th century by Lord Kelvin and enshrined in his Second Law of Thermodynamics. In fact, however, this idea that the natural order must "run down" due to entropy eliminating all sources of "useful work," naively attempts to assign what amounts to permanent and decisive role for negative intentions in particular without formally fixing or defining any role whatsoever for human intentions in general. Whether they arise out of the black-box approach of the mass-balance equation or the unaccounted missing potential energy sources in the energy balance equation, failures in the short-term become especially highly consequential when they are used by those defending the *status quo* to justify anti-Nature "responses", of the kind well-described elsewhere as typical examples of "the roller coaster of the Information Age" (Islam, 2003).

CONCLUSION

This paper presents a comprehensive theory that would form the basis for calculations, required for developing sustainable technologies. It is shown that such calculations cannot be performed with science of tangibles and the notion of intangibles must be introduced. Modern mechanics as well as mathematics are inherently linked to the science of tangibles. In order to solve the problems that include intangibles, a recently developed engineering approach is introduced. It is shown that with this engineering approach, one no longer has to depend on the complex manipulations involving differential equations that are prone to bias the analysis in favor of tangible-only solutions.

REFERENCES

Ball, P., 2007, "Feel the Force", *Nature*, vol. 447, pp. 772-774.

Chhetri, A.B., (2006) Scientific Characterization of Global Energy Sources, *Journal of Nature Science and Sustainable Technology*, vol. 1, No. 3.

Cismondi, M. and Mollerup, J., (2005) Development and Application of a Three-Parameter RK-PR Equation of State, *Fluid Phase Equilibria*, vol. 232, No. 1-2, pp. 74-89.

DDT, (2006) Dichloro-Diphenyl- Trichloroethane from Wikipedia, the Free Encyclopedia [online] Available: (http://en.wikipedia.org/wiki/DDT) [October 4, 2006].

Enron, (2006) Enron Corporation, an American Energy Company from Wikipedia, the Free Encyclopedia [online] Available: (http://en.wikipedia.org/wiki/Enron) [October 4, 2006].

He, J.H., (2005), In search of 9 hidden particles, *Int. Journal of Nonlinear Science and Numerical Simulation*, vol. 6, No. 2, pp. 93-94.

Homeopathy, (2006) Homeopathy from Wikipedia, the Free Encyclopedia [online] Available: (http://en.wikipedia.org/wiki/Homeopathy) [October 4, 2006].

Islam, M.R., (2003) *Revolution in Education* (Halifax NS: EEC Research Group) 552p.

Ketata, C., Satish, M.G. and Islam, M.R., (2006a) The Meaningful Zero, *Proceedings of the 36th International Conference on Computers and Industrial Engineering (ICCIE),* Taipei, Taiwan, R.O.C., June 20-23.

Ketata, C., Satish, M.G., and Islam, M.R., (2006b) The Meaningful Infinity, *Proc. Conference on Computational Intelligence for Modelling, Control and Automation (CIMCA),* IEEE Conference, Sydney, Australia, November 28 – December 1.

Ketata, C., Satish, M.G., and Islam, M.R., (2007a) Abacus-Based Calculators, *Proc. International Conference of Computational and Experimental Engineering and Sciences (ICCES-2007),* Miami, Florida, Jan.

Khan, M.I., (2006) Towards Sustainability in Offshore Oil and Gas Operations, Ph.D. Dissertation, Faculty of Engineering, Dalhousie University, Canada, 440 p.

Khan, M.I. and Islam, M.R., (2007a) *True Sustainability in Technological Development and Natural Resources Management.* Nova Science Publishers, NY, USA, 381 p.

Khan, M.I. and Islam, M.R., (2007b) *Handbook of Sustainable Petroleum Engineering Operations Management,* Gulf Publishing Co., Houston, USA, 452 p.

Lean, G and Shawcross, H., (2007) Are mobile phones wiping out our bees? News, The Independent[online]Available:(http://news.independent.co.uk/environment/wildlife/article2449968.ece).

Maclaren, W., (2007) World fast? Australia switches of incandescent bulbshttp, Treehugger[online]Available:(http//www.treehugger.com/files/2007/02/world_first_aus.php) [15 April, 2007].

Miralai, S., (2006) Replacing Artificial Additives with Natural Alternatives, MASc Thesis, Dalhousie University, Nova Scotia, Canada.

Mittelstaedt, M. (2007), 'Inherently Toxic' Chemical Faces its Future, The Globe and Mail, April 7, pp. A10.

PAD, (2006) Protecting Animals in Democracy [online] Available: (http://www.vote4animals.org.uk/farming.htm) [October 4, 2006].

PVC, (2006) Poly Vinyl Chloride, Wikipedia, the Free Encyclopedia [online] Available: (http://en.wikipedia.org/wiki/Polyvinyl_chloride) [October 4, 2006].

Rey, L, (2003) Thermoluminescence of Ultra-High Dilutions of Lithium Chloride and Sodium Chloride, *Physica* A, vol. 323, pp. 67 – 74.

Schmid, R.E., (2005) Human Pollutants Collect in Bird Droppings, Live Science [online]Available:(http://www.livescience.com/environment/ap_050714_bird_droppings.html) [October 4, 2006].

Soulincode,(2006)[online]Available:(http://www.soulincode.com/images/usa_at_night_nasa.jpg) [October 4, 2006].

Steenhuysen, J., (2007) Mother Nature Still a Rich Source of New Drugs , Environmental News Networks [online] Available: (http://www.enn.com/med.html?id=1442) [April 15, 2007].

Teflon, (2006) Polytetrafluoroethylene from Wikipedia, the Free Encyclopedia [online] Available: (http://en.wikipedia.org/wiki/Teflon) [October 4, 2006].

VIOXX, (2006) Rofecoxib from Wikipedia, the Free Encyclopedia [online] Available: (http://en.wikipedia.org/wiki/Vioxx) [October 4, 2006].

Zatzman, G.M. and Islam, M.R., (2007a) Truth, Consequences and Intentions: The Study of Natural and Anti-Natural Starting Points and Their Implications, *J. Nature Science and Sustainable Technology*, vol. 1, No. 2, pp. 1-38.

Zatzman, G.M. and Islam, M.R., (2007b) *Economics of Intangible*, Nova Science Publishers, New York, p.393.

In: Characterization and Development of Novel Materials ISBN: 978-1-61324-149-3
Editor: Rafiq Islam

Chapter 8

PREPARATION OF STRONTIUM HEXAFERRITE MAGNETS FROM CELESTITE AND BLUE DUST BY MECHANOCHEMICAL ROUTE

R. K. Tiwary,[1] S. P. Narayan[2] and O. P. Pandey[1*]
[1]School of Physics and Materials Science, Thapar University, Patiala, India
[2]Advanced Materials and Process Research Institute (Regional Research Laboratory), Bhopal, India

ABSTRACT

In the present investigation celestite (natural ore of strontium) and blue dust (iron ore fines) have been used for the preparation of strontium hexaferrite powder. The mechanical alloying process has been adopted to prepare strontium hexaferrite powder. The celestite after chemical upradation and physically upgraded blue dust alongwith sodium carbonate was taken for the preparation of strontium hexaferrite in this experiment. The high-energy planetary ball mill with tungsten carbide jar and ball was used to prepare strontium hexaferrite powder. A long time of ball milling for different duration has led to displacement solid-state reaction. At the end of each experiment the product was washed thoroughly and dried. The X-ray diffaction study after annealing shows the development of single-phase strontium hexaferrite after 40 hrs. of milling. The resultant powder was compacted under magnetic field and sintered to prepare the magnet after annealing the ferrite powder. The magnetic properties were measured by Pulse magneto meter. The moderate value of coercivity, remanence and energy product were observed in this sintered magnet. The work illustrates the feasibility to prepare strontium hexaferrite magnetic powders directly from natural ores which can reduce the total cost of production as compared to conventional method.

* Email: oppandey@thapar.edu.

INTRODUCTION

Blue dust is iron ore fines and is found abundantly in the iron ore mines. It cannot be used directly in steel making process due to the fineness of the particle. Celestite is natural ore of strontium and is used for the preparation of strontium carbonate. Strontium hexaferrite is permanent magnet and its formula is ($MFe_{12}O_{19}$) where M stands for Sr, Ba, Pb etc. Its crystal structure is hexagonal magneto plumbite sometimes called M-type ferrite, which is characterized by space group $P6_3/mmc$. [1] It is most widely used permanent magnet, which accounts about 90 wt% of total production of permanent magnets market. Strontium hexaferrite finds its application in motors, loudspeakers, high frequency devices, magnetic recording media etc. [2, 3].

Generally strontium hexaferrite ($SrFe_{12}O_{19}$) is prepared by the conventional method, which involves calcination of the mixture of $SrCO_3$ and Fe_2O_3 in an appropriate ratio at a temperature between 1100-1200^{0}C followed by sintering [4]. The most important factor for the production of $SrFe_{12}O_{19}$ is the cost and availability of the raw materials. Some researchers have tried to reduce the processing cost by using one of the natural raw materials directly. Mortaza Mozaffari et al. [5] have used celestite along with iron oxide to prepare strontium hexaferrite by mechanochemical method. However, they have taken Fe_2O_3 in pure form. Dasgupta et al. [6] used blue dust with pure strontium carbonate for the preparation of strontium hexaferrite by conventional calcination method.

The aim of the present work is to produce strontium hexaferrite powder directly from celestite and blue dust which are natural ore of strontium and iron by mechanochemical method with a view to make the process cost effective and to explore the possibility of commercial production of strontium hexaferrite directly from the ores. This will also help to utilize the waste and discarded fines that are causing environmental problem.

EXPERIMENTAL PROCEDURE

The starting raw materials taken for this study are celestite, blue dust and sodium carbonate. The celestite ore was obtained from Trichirrapalli district of Tamil Nadu (India), which was supplied by M/s J.M. Mines and Minerals, 37, Williams Road, Cantonment, Tamil Nadu (India). The ore was ground to fine powder in a mortar and pestle. The powder was sieved.

Table 1. Chemical analysis of celestite before and after acid treatment

Compound	Before acid leaching (wt %)	After acid leaching (wt %)
SrSO4	91.60	96.66
$SrCO_3$	1.30	<0.02
$BaSO_4$	0.90	0.87
Fe_2O_3	4.01	0.60
SiO_2	0.10	0.10
Al_2O_3	0.21	0.15
CaO	0.36	0.01

Table 2. AAS results for other trace elements before and after acid treatment of celestite

Element	Before acid leaching(wt %)	After acid leaching(wt %)
Mg	0.12	0.012
Pb	0.24	0.021
Zn	0.01	0.01
Ni	0.006	0.005
Cu	0.29	0.002
K	0.026	0.021
Na	0.24	0.002

The fraction less than 45 micron was collected for the experimental work. The chemical analysis of celestite and blue dust was carried out to know its chemical composition. For different trace elements analysis, atomic absorption spectrometer (flame type model no. GBC-932AA, Australia) was used.

Chemical treatment of celestite ore: Celestite was chemically treated to remove gangue materials. For it 100 gm celestite powder (<45μm) was taken in 2 liter borosilicate glass beaker. About 500 ml of 1:1 HCl, 20 ml conc. HNO_3 and small amount of NH_4 F was added in this fine powder mass. All acids and other chemicals were of laboratory grade purity. The beaker containing acid and celestite was heated on a hotplate at 60°C temperature with constant stirring for long duration (36 hours). The acid portion was decanted and fresh HCl (1:1) was added followed by small amount of HNO_3 addition to repeat this process. Acid treated celestite was filtered and washed many times with distilled water till the pH of residue on the filter paper attains neutral value. The filter paper with residue was dried in an oven at 110°C for 2 hours. By addition of HCl and HNO_3 the strontium carbonate present with celestite along with Fe, Ca, dissolves and remains in the dissolved condition in the liquid. NH_4F was added to dissociate any silicate materials present in celestite. By chemical treatment process celestite upgradation is performed and acid soluble impurities like Fe, Ca, etc. get reduced significantly. The chemical composition of celestite before and after acid treatment is given in table 1 and 2 respectively.

UPGRADATION OF BLUE DUST

Blue dust was collected from Bailadila mines (Chattishgarh state, India). The particle size <35 μm was collected for the upgradation purpose. For it 100 gm of blue dust was taken in a 2 liter beaker. Beaker containing blue dust was heated on a hot plate at 70°C and compressed air was passed in it through indigenously designed glass nozzle.

This leads to continuous stirring of blue dust particles. After half an hour heating and blowing process the water was discarded. This process was repeated five times to ensure complete removal of water soluble impurities. After final washing the mass was filtered and dried in an oven at 110°C for 1hour. The schematic view of the froath floatation setup for the upgradation of blue dust fine particles is shown in figure 1. The black particles represent the blue dust fines and grey circles the air bubbles.

Preparation of Strontium Hexaferrite Powder

The appropriate quantity of treated celestite, upgraded blue dust (mole ratio SrO/Fe_2O_3 =5.5) and sodium carbonate with 30% excess amount to stoichiometic ratio was taken. A batch of 50 gm. mixture was mixed using zircon ball and vessel in a high-energy planetary mill (Model Mega pact, Pilamec Ltd. U.K.) in a medium of analytical grade reagent (AR) hexane.

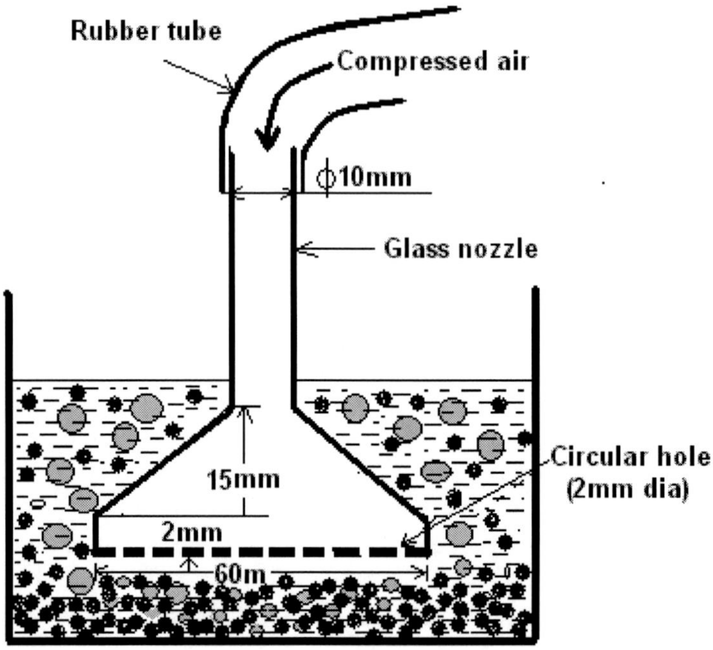

Figure 1. Froth floatation setup for blue dust upgradation.

After 1 hour of mixing the material was taken out and put into tungsten carbide jar with tungsten carbide balls. The charge to ball ratio was fixed (1:10). Hexane was used as medium for mixing which continued up to 50 hours. At every 10 hours interval sample colour was monitored till the mixture colour turned to black.

During initial experimentation hexane was added to get homogenized mass. As the milling proceeds hexane evaporates leaving fine powders. These dried powders undergo impact welding leading to diffusion in each other of the constituent ingredients in solid state. The mixture was taken out after milling. The mixture was characterized with XRD to see the phases present in the product mass. It was observed that formation of ferrite phase does not start in the begining after 1 hour mixing as can be seen in figure 2(a). However, distortion in peaks was observed after 30 hrs of ball milling. Finally after 40 to 50 hrs. of ball milling the powder was annealed at 900^0C for 1 hour in a muffle furnace to see the presence of phases. Once the phase of strontium hexaferrite after 50 hours of ball milling followed by annealing was confirmed then these powders were selected to prepare strontium hexaferrite sintered magnets. For it 5% PVA (polyvinyl alcohol) as a binder was added to this powder and mixed well to make slurry and was allowed to dry in the open air. The pellet of 5mm diameter and 11 mm length was prepared by applying compressive force of 500 MPa under magnetic field of 2 T to make it anisotropic. The pellet was sintered at 1250^{0}C for an hour in a tubular furnace in open air.

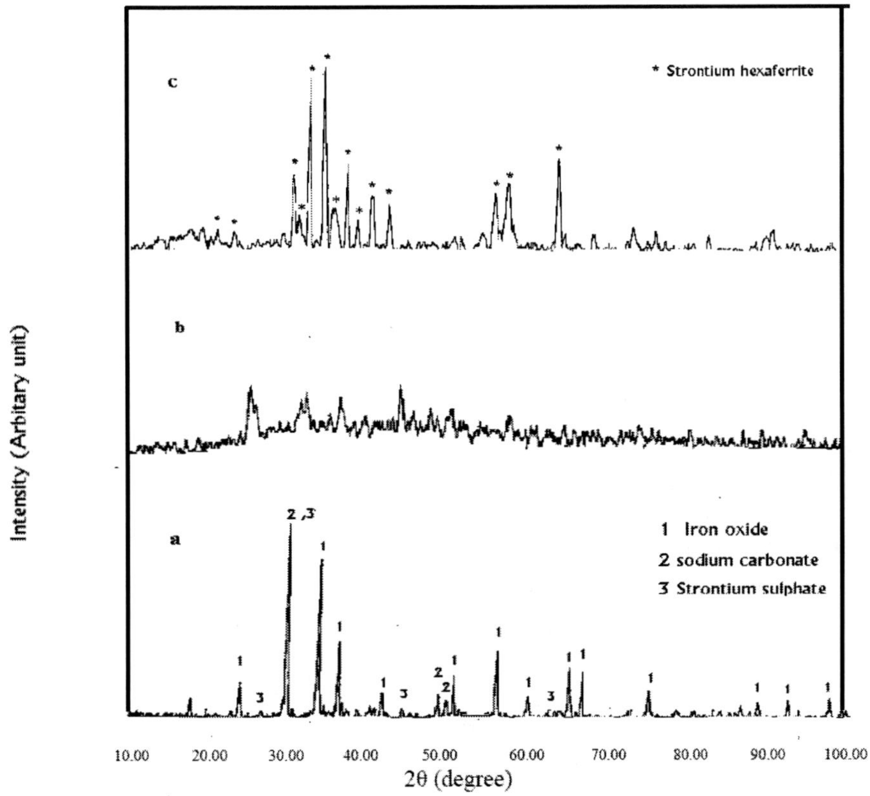

Figure 2. X-ray diffraction pattern of the powders mechanically alloyed for various times (a) after 1 hr. milling (b) after 30 hrs. of milling (c) after 50 hrs. of milling followed by annealing.

RESULTS AND DISCUSSION

The celestite powder was chemically treated for its upgradation. The percentage value of different ingredients before and after acid leaching is given in table 1. It was observed that even after acid treatment some impurities like Fe_2O_3, $BaSO_4$, SiO_2, Al_2O_3 etc. remain in small quantity. The presence of these compounds is not favorable for the magnetic properties of strontium hexaferrite produced there of. The amount of trace element present in celestite before and after leaching is given in table 2. The chemical analysis and trace analysis of source and upgraded blue dust is given in table 3and 4.

The reaction involved in milling can be written as:

$$SrSO_4 + Na_2CO_3 + 5.5\ Fe_2O_3\ SrFe_{12}O_{19} + Na_2SO_4 + 3O_2$$

The characteristic feature of solid-state reaction through mechanical alloying is that it involves the formation of product phases at the interfaces of the reactants.

Table 3. chemical analysis of blue dust before and after upgadation (floatation)

Compound	Before upradation (wt %)	After upgradation (wt %)
Fe2O3	96.2	97.54
Al2O3	0.322	0.26
SiO2	1.60	1.21
CaO	0.94	0.13

Table 4. Trace elemta analysis of blue dust (AAS) before and after upgradation

Element	Before upgradation (wt %)	After upgradation (wt %)
Mg	0.40	0.012
Pb	0.027	0.021
Zn	0.01	0.01
Cu	ND	ND
K	0.026	0.021
Na	0.24	0.002

The growth of the product phases occur by diffusion of atoms of the reactant phases through the product phases, which creates a layer preventing further reaction.

Intensive milling increases the area of contact between the reactants powder particles due to reduction in particle size which allows fresh surfaces to come into contact. This allows the reaction to proceed without the necessity for diffusion through the product layer due to which solid state reaction that require high temperatures will take place at lower temperature during mechanochemical synthesis without any heat treatment. It is well known that conventional process for the preparation of $SrFe_{12}O_{19}$ is through solid-state reaction at high temperature. This requires lot of chemicals of high purity. Apart from this the requirement for calcination, milling, high sintering temperature makes the process costly. The advantage of mechanical milling, process over other conventional method is that it does not require any external heating source for solid-state reaction to occur [7-12].

Phase Formation

The X-ray study for the formation of strontium hexaferrite ($SrFe_{12}O_{19}$) phase is presented in figure 2 (a-c). The XRD pattern after 1 hour of ball milling of the mixture of celestite, sodium carbonate and iron oxide show the presence of all constituent phases which is shown in figure 2 (a). The figure 2 (b) represents the X-ray diffraction pattern of initial mixture after 30 hrs. of mechanical milling. It indicates the formation of strontium hexaferrite phase. However, presence of ingredient phases also exist. The much weak and broaden peaks of Fe_2O_3 and $SrSO_4$ are visible with formation of small amount of amorphous $SrFe_{12}O_{19}$ phase [13-15]. The peak broadening of initial oxides indicates that shorter time of mechanical activation is not sufficient to obtain crystalline phase of $SrFe_{12}O_{19}$. However, similar pattern was also observed even for higher mechanical milling time indicating that reaction is not completed. The figure 2 (c) shows the typical XRD pattern after 50 hrs. of mechanical milling followed by annealing at 900^0C for 1 hour. Here we can find that there is decrease in amorphization nature. More sharp peaks of $SrFe_{12}O_{19}$ are visible in this pattern showing the increased volume of ferrite phase which has formed after 50 hrs. of milling followed by annealing. The X-ray pattern shows the formation of single-phase strontium hexaferrite phase. The particle size of $SrFe_{12}0_{19}$ powder after annealing was measured by applying advance Scherrer equation. The calculated value of ferrite particle is 17.1nm. It appears that even after annealing the particles are of nano size which after sintering get converted to micron size.

The lattice parameter was calculated using following formula.

$$(1/d)^2 = 4/3\ [\ (h^2+hk+k^2)/a^2\ ] + l^2/c^2$$

where d = inter spacing between planes
hkl = Miller indices
c, a = lattice parameter.

The theoretical values of c and a are 23.037 and 5.886 Å whereas calculated values from X-ray diffractogram of these are 23.289 and 5.6613 Å respectively. The observed lattice parameters are close to the theoretical values. Magnetic properties of ferrite powder produced after different milling time followed by annealing is shown in figure 3. By increasing the milling time magnetic properties are observed to increase. This is due to increasing volume of strontium hexaferrite phase. This is because of the fact that annealing releases the mechanical stresses existing in powders due to impact load [16]. The typical B-H loop of the sintered anisotropic strontium hexaferrite is shown in figure 4.

Magnetic Properties

Mortaza Mozaffari et al. [5] have reported the direct utilization of celestite for the preparation of strontium hexaferrite powder using high purity iron oxide after longer time of milling (24 hours). They have employed milling process to achieve intimate mixture of celestite, sodium carbonate and iron oxide. The milled powder was calcined at 1000^0C for 10 hours. In their process no mechanochemical reaction has taken place. The product phase (strontium hexaferrite) was obtained after calcination. The magnetic properties of these sintered sample is shown in table 5.

Miclea et al. [13] have produced strontium hexaferrite powder by mechanical alloying method. However, they have used strontium carbonate and iron oxide of high purity. Severe stresses and structural deformation was reported in their work. These stresses were relieved by annealing process. Strontium hexaferrite prepared with high purity raw materials exhibit good magnetic properties (Table 5). The coercivity is the extrinsic property, which depends mainly on particle size, annealing temperature and packing factor of the material.

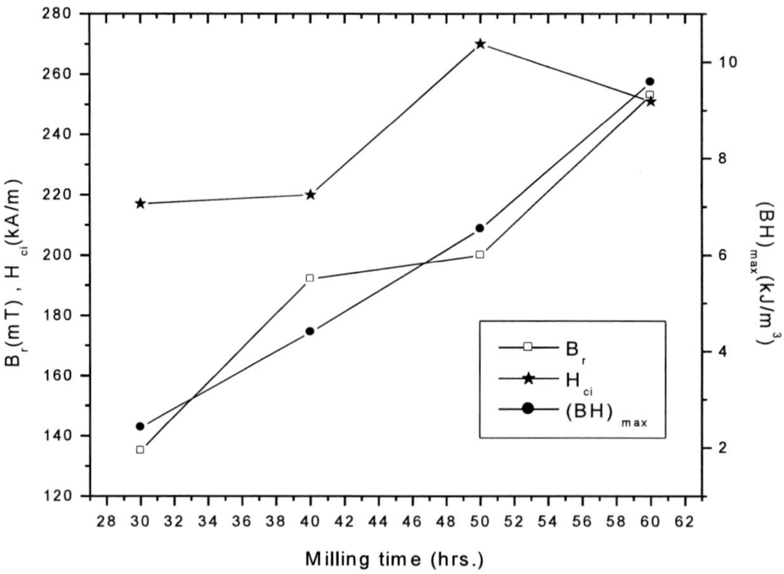

Figure 3. Variation of B_r, H_{ci} with different milling time after annealing and sintering.

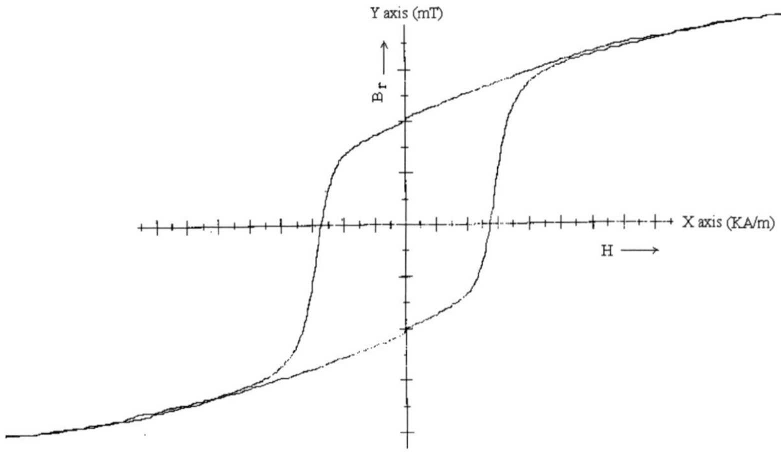

Figure 4. B-H loop of anisotropic $SrFe_{12}O_{19}$ sintered magnet.

Table 5. Comparision of magnetic properties of strontium hexaferrite by mechanochemical alloying

Sr.No.	Ingredients	Magnetic properties References B_r (Gs)	H_{ci}(Oe)	BH_{max} (MGOe)	
1	Celestite, sodium carbonate iron oxide (pure)	2305	5230	1.2	Mortaza Mozaffari et al.[5]
2	Strontium carbonate (pure), iron oxide (pure)	2100	4600	1.85	C. Miclea et al.[13]
3	Strontium hexaferrite, iron sulphide	-	5700	-	J.Sort et al.[14]
4	Strontium hexaferrite	2400	4200	1.20	S.V.Ketov et al [15]
5	Celestite, sodium carbonate blue dust	2533	3160	1.19	Present investigation

The single domain size of strontium hexaferrite is about 1μm. [10]. Due to mechanical alloying particle size has reduced to nano size. Sort et al. [14] in their experiment tried to improve the coercivity of the strontium hexaferrite by ball milling of pure $SrFe_{12}O_{19}$ and FeS. The coercivity enhancement observed is due to the formation of α-Fe_2O_3 phase which has formed after ball milling (table 5). S.V.Ketov et al. [15] have made the nanocrystalline $SrFe_{12}O_{19}$ powder using high energy milling process from pure strontium hexaferrite powder. This process covers only improvement in magnetic properties by high energy milling and annealing (table 5).

The present study involves the solid-state reaction by mechanochemical reaction using celestite and blue dust. After the reaction the powder was annealed. The magnetic properties show the moderate value of remenance and energy product and the coecivity as shown in table 5.

The chemically upgraded celestite and modified blue dust has been used in the present experiment. Some siliceous impurities are always occluded with iron ore, which in all proportion does not favour the magnetic properties. The celestite even after chemical upgradation is not free from impurities. The lower value of magnetic properties may be attributed due to presence of retained impurities as discussed above. Apart from this the value of sintered density of the magnet obtained in this experiment is 4.9 gm/cc, which is lower than the theoretical density 5.1 gm/cc [16]. The decrease in density also causes to lower $(BH)_{max}$ value. However, the magnetic properties obtained in the present investigation are within the range of other researchers as given in table 5 where they have used milling process to synthesize strontium hexaferrite.

The prolonged milling led to the decrease in the value of B_r, H_{ci} and $(BH)_{max}$ due to severe mechanical stresses and by formation of soft magnetic amorphous phase. After annealing the magnetic properties increases significantly due to recrystallisation [16-21].

Microstructural Studies

Figure 5. SEM micrograph of $SrFe_{12}O_{19}$ powder after annealing.

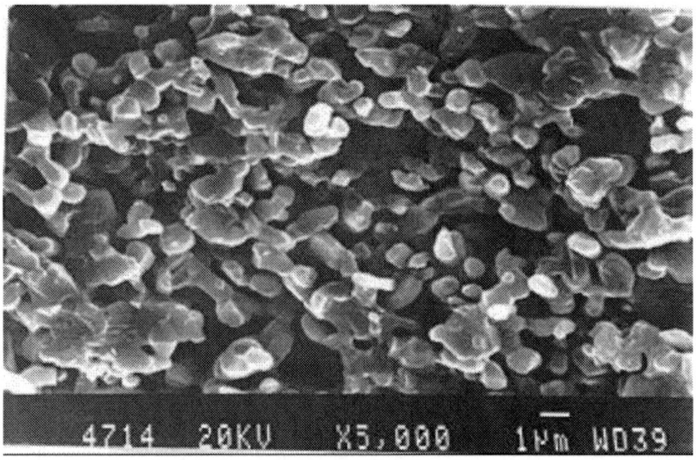

Figure 6. SEM of fractured surface of sintered $SrFe_{12}O_{19}$ magnet.

Figure 5 shows the SEM micrograph of strontium hexaferrite annealed powder, which was obtained after gold sputtering. This shows the hexagonal grains of ferrite. Figure 6 shows SEM micrograph of the fractured surface of the ferrite magnet. The major portion of particles is in range of 1 μm. The grains are elongated showing orientation of ferrite particles. During the process of annealing and sintering small particles of strontium hexaferrite have grown to single domain ≤1 μm.

CONCLUSION

The result of magnetic properties obtained in the present investigation shows that strontium hexaferrite with moderate values of remenence, coercivity and energy product can be obtained by mechanical alloying of celestite and blue dust followed by annealing. These magnets can be used in those components where moderate magnetic properties are required. The most important features of this work are as follows.

(a) The lower cost of starting raw minerals (celestite and blue dust) will make the process more economical to produce strontium hexaferrite magnets as compared to the other conventional methods where high purity chemicals are utilized.
(b) The mechanochemical method is direct method for the solid-state reaction and this does not require any additional step like conversion of celestite to strontium carbonate and high temperature reaction. Therefore it makes the process easy and economical.
(c) This innovative process will definitely help to exploit the large deposit of Indian celestite and waste blue dust on commercial scale for their utilization and for the development of strontium hexaferrite, which exhibit moderate magnetic properties and is cost effective.
(d) The values of magnetic properties obtained in present work are within the range of those reported using mechanochemical work as has been compared in table 5.

ACKNOWLEDGMENT

The authors are thankful to Department of Mines, New Delhi (INDIA) for providing necessary funds for this work.

REFERENCES

[1] Krik-othmer, (1993), "*Encyclopedia of chemical technology*" fourth edition, pp 393-405, John Willey and Sons.

[2] Joyce A. Ober. (2007), Strontium USGS Report, pp 1-7.

[3] J Pual, Mac Millan, (1994), "*Strontium and Strontium compound*" Ullmann's Encyclopedia of Industrial Chemistry, A-25.

[4] Puneet Sharma, Amitabh Verma, R.K.Sidhu, O.P.Pandey, (2005), "*Process parameter selection for strontium ferrite sintered magnets using Taguchi L9 orthogonal design*" Journal of Material Processing Technology, vol.168, pp 147-151.

[5] Mortaza Mozaffari, Jamshid Amighian, (2002), "*Direct use of celestite to prepare presintered* $SrFe_{12}O_{19}$ *powders*" Physica B, vol. 321, pp 45-47.

[6] R. Dasgupta, S.K. Bose, S.P.Narayan, (1995), " *Hard ferritess from blue dust of Bailadila mines of Madhya Pradesh*" Research and Industries, vol. 40, pp 236-239.

[7] G.R. Karagegedov, N.Z. Lyakhov, (2003), "*Mechanochemical grinding of inorganic oxides*" KONA, vol. 21, pp 76-87.

[8] V. V., Boldrev, (1996), "*Mechanochemistry and mechanical activation*" Materials Science Forum, vol. 227, pp 511-520.

[9] E. Gaffet, D. Michael, L. Muzerolles and P. Berther, (1997), "*Effect of high- energy ball milling on ceramic oxide*" Matterials Science Forum, vol. 235-238(1), pp 511- 520.

[10] C. Suryanarayna, (1995), "*Nanocrystalline materials*" International Materials Reviews, vol.40, No.2, pp 41-115.

[11] E. Gaffet, F. Bernard, J.C. Niepce, F. Charlot, C.Gra, G. Caer, (1999), "*Some recent developments in mechanical activation and mechanosynthesis*" J. Material Chemistry, vol. 9, pp 305-314.

[12] E. Gaffet, D. Michael, L. Muzerolles and P. Berther, (1997), "*Effect of high- energy ball milling on ceramic oxide*" Matterials Science Forum, vol. 235-238(1), and pp 511-520.

[13] C. Miclea, C. Tanasoiu, C.F. Miclea, I. Spanulescu, A. Gheorghiu and M. Cioangher, (2006), "*Structure and magnetic properties of strontium hexaferrite nanopowders prepared by mechanochemical synthesis*" Advance in Science and Technology, pp 321-326.

[14] J. Sort, J. Noguess, S. Surifiach, J.S. Munoz, M.D. Baro, (2003), "*Coercivity enhancement in ball-milled and heat treated Sr-ferrite with iron sulphide*" Journal of Metastable and Nanocrystalline Materials, vol. 15-16, pp 599 –606.

[15] S.V.Ketov, Yu. D. Yagodkin, A.L.Lebed, Yu. V.Chernopyatova, K. Kholpkov, (2006), "Structure *and magnetic properties of nanocrystalline* $SrFe_{12}O_{19}$ *alloy produced by high-energy ball milling and annealing*" Journal of Magnetism and Magnetic Materials, vol. 300, pp e479-e481.

[16] H.Taguchi, F. Hirata, T. Takeishi, (2000), United State Patent: 6,132,635, pp 1-24.

[17] B.D. Cullity, (1972),"*Introduction to magnetic materials*" Addition Wasley, Reading, MA.

[18] W. A. Kaczmarek, S.J.Campebell, E.Wu. B.Idrikowski, K.H.Miller, (1998), "*Structural and magnetic phases in ball milled strontium ferrite*" Material Science Forum, vol. 269-272, pp 465-460.

[19] Spomenka Besenicar, Miha Drofenic, (1991), "*High coercivity Sr hexaferrites*" Journal of Magnetism and Magnetic Materials, vol. 101, pp 307- 309.

[20] P. I. Paulin Filno, R.R.Correa, (2005), " *Chatecterisation of strontium ferrite powders obtained by high energy milling*" Material Science Forum, vol. 498-499, pp 311-315.

[21] S. P. Gubin, Yu A Kokasharow, G.B. Khomutov, (2005), "*Magnetic nanoparticles, preparation, structure and properties*" Material Science Forum, vol. 74, pp 489-520.

In: Characterization and Development of Novel Materials
Editor: Rafiq Islam
ISBN: 978-1-61324-149-3

Chapter 9

AN IMPROVED HPTLC METHOD FOR DETERMINATION OF SENNOSIDES IN *CASSIA ANGUSTIFOLIA*

Wasim Aktar[1*], Rajesh Muthukumar[2], M. Paramasivam[1] and Dwaipayan Sengupta[3]

[1]Department of Agricultural Chemicals, Bidhan Chandra Krishi Viswavidyalaya, Mohanpur, Nadia, W.B, India
[2]Department of Spices and Plantation Crop, Faculty of Horticulture, Tamilnadu Agriculture University, Koimbatore, T.N., India
[3]Department of Agricultural Chemistry and Soil Science, Institute of Agricultural Science, University of Calcutta, Kolkata, India

ABSTRACT

A simple high-performance thin-layer chromatographic (HPTLC) method has been developed for the simultaneous determination of the pharmacologically important two hydroxy anthracene glycosides *viz.* sennoside A and sennoside B in *Cassia angustifolia*. The assay combines the separation and quantification of the analytes on silica gel 60 GF_{254} HPTLC plates with visualisation under UV and scanning at 350 nm. Using this technique, the alkaloidal content of different parts of the title plant have been determined.

Keywords: HPTLC; quantitative determination; sennoside A; sennoside B; Cassia angustifolia; Caesalpiniaceae.

---

* Correspondence to : wasim04101981@yahoo.co.in.

1. Introduction

Cassia angustifolia (Caesalpiniaceae), commonly called Senna, is a well-known plant drug in Ayurvedic and Unani medicine (Kapoor, 2001; Atal et al, 1982; Das et.al., 2003 and Maetindale,1977). It has been used for the treatment of various diseases and disorders particularly for the treatment of constipation (Sharma, 2004). Senna is a strong purgative that should be taken in proper dosage otherwise it may lead to gripping and colon problem (Bhattacharjee,2004).

Sennoside A (**1**) and sennoside B (**2**), two major pharmacologically important anthracene glycosides, have been isolated from *C. angustifolia* (Evans *et al,* 1979) and has been shown to possess purgative properties. Methods so far available for the determination of these alkaloids are very cumbersome and time-consuming and also not economically viable.

Therefore it was thought worthwhile to develop a simple and high-precision HPTLC method for simultaneous analysis of **1** and **2** occurring in the pods, leaves, stems and roots of *C. angustifolia*.

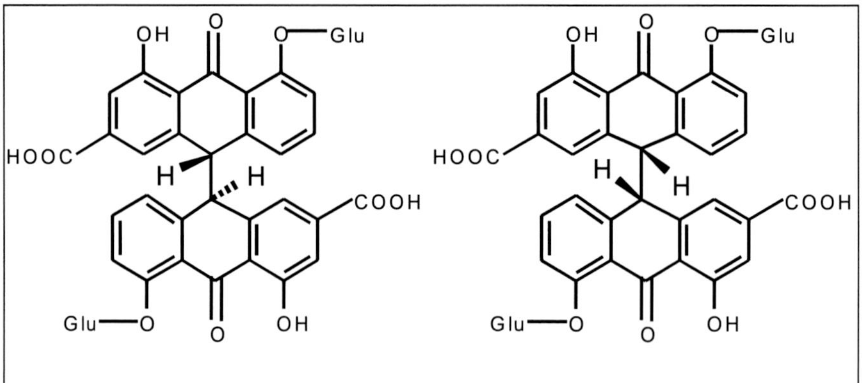

Sennoside A Sennoside B

2. Materials and Methods

2.1. Plant Material

Pods, leaves, stems and roots of *C. angustifolia* were collected from the Experimental Farm for the Medicinal and Aromatic Plants (TNU, India). The plant specimens were authenticated and a voucher specimen is deposited in the herbarium.

2.2. Chemicals and Standard Alkaloids

Reagents used were from Merck (Darmstadt. Germany). In order to obtain standards of **1** and **2**, dried powdered pods (1 kg) of *C. angustifolia* were extracted with ethanol (Openshaw, 1953) and an alkaloid fraction (2.4 g) isolated. This fraction was subjected to column chromatography over silica gel (Merck) using mixtures of chloroform and methanol of increasing polarity. Sennoside A (**1**) was eluted from the column with chloroform: methanol

(49:1) as a yellowish solid and was crystallised from chloroform: ethyl acetate. Sennoside B (**2**) was eluted from the Column with chloroform: methanol (9: 1) and crystallised from ethanol. The identities of **1** and **2** were confirmed by comparison of their spectral data with those previously reported (Franz,1993;Terreaux et.al.,2002 and Yiping et.al.,2007).

2.3. Extraction of Plant Material for Analysis

Air dried (30-50°C) powdered pods, leaves, stems and roots (each 1.0 g) of *C. angustifolia* were extracted separately with 3x20 mL methanol containing 20% (v/v) ammonium hydroxide. Extracts were concentrated under vacuum, redissolved in methanol, filtered, reconcentrated and finally made up to 20 mL volume with methanol prior to HPTLC analysis (Terreaux et.al.,2002).

2.4. Chromatographic Conditions

Chromatography was performed on glass-backed silica gel $60F_{254}$ HPTLC layers (20x20 cm, 300 µm layer thickness) prepared using a Camag (Multenz, Switzerland) TLC plate auto-coater. Methanolic solutions of samples and standard compounds **1** and **2** of known concentrations were applied to the layers as 8 mm-wide bands positioned 15 mm from the bottom and 20 mm from the side of the plate, using a Camag Linomat 5 automated TLC applicator with the nitrogen flow providing a delivery speed of 150 nL/s from the syringe. These parameters were kept constant throughout the analysis of samples.

2.5. Detection and Quantification of the Alkaloids

After completion of sample application, the plate was developed in a Camag Twin through glass tank presaturated with mobile phase of 2-propanol: ethyl acetate: water: Formic acid (17:19:12:02) for one hour. The TLC runs were performed under laboratory conditions of 25±2°C and 60% relative humidity. After development the plates were taken off and dried by drier. Sennosides A and B were quantified using a Camag TLC Scanner model 3 equipped with Camag Wincats software applying the following conditions: slit width 6x0.45mm, wave length (λ_{max}) 350nm, absorption-reflection scan mode. The identification of Sennoside A and B in formulations was confirmed by superimposing the UV spectra of samples and standards within the same R_f window. In order to prepare calibration curves, stock solutions of **1** and **2** (1 mg/5 mL each) were prepared and various volumes of these solutions were analysed by HPTLC exactly as described above. Calibration curves of peak area vs. concentration were prepared.

3. RESULTS AND DISCUSSION

Different compositions of the mobile phase for HPTLC analysis were tested in order to obtain high resolution and reproducible peaks. The desired aim was achieved using 2-propanol: ethyl acetate: water: Formic acid (17:19:12:02) as the mobile phase. The wavelength of 350 nm was found to be optimal for the highest sensitivity (Figure 1). The calibration curves for the alkaloids sennoside A (**1**) and sennoside B (**2**) were linear in the range 100-1000 ηg (Table 1).

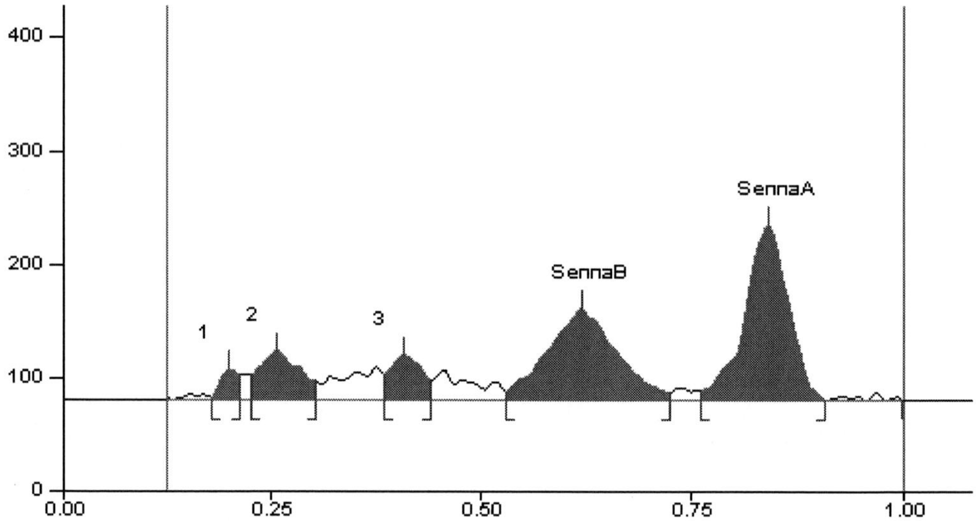

Figure 1. Scan (at 350nm) showing the separation of sennoside A and B in the extract of a laxative formulation.

Table 1. R_f values by HPTLC and linear regression equations for the determination of sennoside A and sennoside B

Compound	R_f value	Regression equation	*r*
Sennoside A (1)	0.63	Y= 1631.9 x + 6840.5	0.987
Sennoside B (2)	0.84	Y= 5676.9 x + 1605.7	0.997

The accuracy of the determination of the recovery rate was determined by triplicate analyses of the plant samples spiked with three different concentrations of stock solution of **1** and **2**. The recovery rates were 98% for **1**and 93% for **2**.

For the quantitative determination of sennoside A (**1**) and sennoside B (**2**), the analyses of different parts of specimens of *Cassia angustifolia* were repeated three times. The average content of **1** and **2** in the pods, leaves, stems and roots are given in Table 2. It is clear that both alkaloids were present in each plant part having, their maximum concentrations in the pods.

Table 2. Distribution of the alkaloids sennoside A and sennoside B in different parts of *Cassia angustifolia*

Plant parts	Sennoside A (1) (%dry weight)[a]	Sennoside B (2) (% dry weight) [a]
Pods	0.97	0.143
Leaf	0.65	0.047
Stem	0.31	0.027
Root	0.062	0.0013

[a] Mean values (n = 3).

CONCLUSION

In summary, the HPTLC method for the simultaneous analysis of sennoside A and sennoside B from *C. angustifolia* as well as from various commercial formulations for quality evaluation, reported here is very simple, sensitive, economic and suitable for rapid screening of large number of plant samples. This analysis can be performed without any special sample pre-treatment and fifteen samples can be analysed on a single TLC laver (20 x 20 cm). The method is less expensive than an LC method. The method is, thus, suitable for quality control laboratories, where economical use of time and space is essential.

ACKNOWLEDGMENT

Authors are very much grateful to Department of Biochemistry, TNU and Department of Applied Biochemistry, Rajabajar Science College, University of Calcutta for providing instrumental facilities in carrying out this research work.

REFERENCES

Atal CK, Kapoor BM 1982. *Cultivation and utilization of Medicinal plants*, RRL, Jammu Twai, India. P-8.

Bhattacharjee SK 2004. A hand book of Medicinal plant, Pointer Publisher,Jaipur 74-79.

Chopra RN, Nayar SL, Chopra IC. 1956. *Glossary of Indian Medicinal Plants.* CSTR: New Delhi; 7.

Das PN, Purohit SS, Sharma AK,Kumar T 2003. *A handbook of Medicinal plants*, Agrobios(India), Jodhpur,118.

Evans FJ, Lee MG, Games DE 1979. Electron impact, chemical ionization and field desorption mass spectra of some anthraquinone and anthrone derivatives of plant origin. *Biological Mass Spectrometry*. 6(9): 374-380.

Franz G. 1993. The Senna Drug and Its Chemistry. *Pharmacology*. 47 :2-6.

Habib AA, EI-Sebakhy NA 1980. Spectrophotometric estimation of sennosides and rhein glycosides in senna and its preparations. *J. Nat. Prod.* 43(4): 452-458.

Kapoor LD. 2001. *Handbook of Ayurvedic Medicinal Plants.* CRC Press: Boca Raton, FL; 216.

Khafagy SM, Girgis AN, Khayyal SE, Helmi MA 1972. Estimation of sennosides A, B, Cand D in senna leaves, pods and formulation. *Planta Med.* 21(3):304-309.

Lin YT, Huang CY 1998. Determination of sennoside A and B in Diet tea by HPLC. *J. Fd. and Drug Anal.* 6(1):433-438.

Maetindale 1977. *The extra Pharmacopoeia*, 27th ed, the pharmaceutical press, London, 1342-1343.

Openshaw HT. 1953. The quinazoline alkaloids. In *The Alkaloids, Vol.* 3, Manske RHF, Homes HL (eds). Academic Press: New York; 101-118.

Shah SA, Ravisankara MN,Nirmal A, Shishoo Cj, Rathod Is, Suhagia BN. 2000. Estimation of individual sennosides in plant material and marketed formulations by an HPTLC method. *J. Pharm. Pharmoco.* 52: 445-449.

Sharma R 2004. *Agro-Techniques of Medicinal plants*, Daya publishing House, Delhi, 176-177.

Terreaux C, Wang Q, Robert JI, Ndjoko K , Grimminge W, Hostettmann K. 2002. Complete LC/MS Analysis of a Tinnevelli Senna Pod Extract and Subsequent Isolation and Identification of Two New Benzophenone Glucosides. *Planta Med.* 68: 349-354.

Yiping L, Zhen H, Langchong H. 2007. An approach to develop binary chromatographic fingerprints of the total alkaloids from Caulophyllum robustum by high performance liquid chromatography/diode array detector and gas chromatography/mass spectrometry. *Journal of Pharmaceutical and Biomedical Analysis.* 43(5):1667-1672.

In: Characterization and Development of Novel Materials
Editor: Rafiq Islam
ISBN: 978-1-61324-149-3

Chapter 10

INVESTIGATION OF THE PROPERTIES OF A BOROSILICATE GLASS FROM RECYCLED MANGANESE CRUST TAILINGS

G. F. T. Lay,* M. C. Rockwell[1] and J. C. Wiltshire[2]
[1]Civil and Resource Engineering, Dalhousie University,
Halifax, Nova Scotia, Canada
[2]Hawaii Undersea Research Institute (HURL), MSB 303;
Honolulu, HI

ABSTRACT

Manganese crust tailings vitrification with boric acid as flux and silica as an additive at 1160C was aimed at recycling waste tailings into borosilicate glass materials. Glass products characterization and testing was conducted using TGA, DTA SEM/EDX and optical microscope and revealed acceptable physical and chemical characteristics. Glass 5-15 exhibited a density of 2.3 g/m^3, H_v of 4.8 GPa, K_c of 0.66 MPa $m^{1/2}$, E value of 92, TEC of 2.89 in x 10^{-6} and acceptable chemical resistance. The glass was able to be drawn into fibers of 200 to 400 micron diameter, which suggests its potential use in producing high-end marketable products and indicates the possibility of future research into additional applications.

Keywords: products characterization, manganese crust tailings, borosilicate glass, waste, vitrification, recycle, glass properties, tailings management, glass fiber.

* tlay@dal.ca, maria.rockwell@dal.ca.

1. Introduction

The chief aim of vitrification is to produce a homogeneous glass possessing a high chemical durability (Frugifer et al, 2002). To be employed as an industrial product, the glass should also exhibit acceptable physical characteristics. The chief advantage of vitrification is that glass is the material, which possesses the highest chemical stability and it can incorporate into its structure, virtually any element of the periodic table. This is important, since wastes have a very complex chemical composition (Scarinci et al, 2000).

Shelby (2005) suggested that acceptable quality glasses usually require sufficient time for the stirring action of rising bubbles and mechanical stirring action to reduce heterogeneity in the melt. This additional time and stirring action is usually not available for lab-tests, thus glasses produced in the lab are much more prone to developing flaws. Volatilization of batch components such as Fe, B and Mn oxides may assist stirring action by releasing O_2 and creating bubbles, but flaws such as straie and chord can be formed by the melt-atmosphere boundary. Sheng et al. (2003) suggested that to reduce volatility of metal oxides, the typical melting temperature employed should be <1200C, with melt viscosity being <100 Pa s, such that melt pouring is relatively easy.

Gelhoff and Thomas (1926) and Morey (1954) found that substituting B_2O_3 for SiO_2 increased the elastic modulus and strength of experimental glasses substantially more than substitution of Al_2O_3 on a wt% basis. Smith (1993) noted that when boric oxide enters the silica network, it weakens the structure and lowers the melting point of the silica glass. Scarinci et al. (2000) reported a decreasing TEC value with increasing waste content, probably due to a decrease in Na_2O in the resulting glasses.

Chemical durability of waste glasses is directly related to the extent these glasses resist chemical reactions with water. Constant exposure of alkali glasses, for instance, can cause extraction of alkali and an increase in the pH of the water (Rawson,1991). In addition, waste glasses undergo a variety of complex changes in aqueous environments, which are referred to as glass corrosion or glass dissolution (Cunnane et al.,1992)(Feng et al,1993). Ballard and Pye (1976) reported that MnO and Fe_2O_3 are fairly alkaline-resistant and have a fairly large glass formation region in silicate systems. Paul and Youssefi (1978) reported that molar substitution of FeO and MnO for CaO in CaO-Na_2O-SiO_2 glasses increased the alkaline durability. In addition, it has been suggested that when glass crystallizes, while the glass block is being cooled, glass properties such as chemical durability could change dramatically (Luckscheiter and Nesovic (1996).

Sheng (1997) determined that chemically durable glasses usually also exhibit high mechanical strength, while Besmann and Spear (2002) reported that crystal formation and phase separation can greatly affect leaching behaviour in glasses. Shelby (2005) reported that durability of phase separated glass is always lower than that of homogeneous glasses, while degradation of chemical durability can be offset by adding modifiers such as Al_2O_3, alkaline earth oxides (Ca, Mg, Ba), or transition elements (Mn, Fe, Cu) as oxides to the glass material.

Previous authors have described various methods for producing glass fibers and beads rock wool, glass foams and glass matrix composites from waste materials, while Hood and Norberg (1984) explained the ability of sodium borosilicate glasses to be leached to yield porous glasses during their work to patent Vycor. Glass fibers are used as reinforcing fibers, chemical fibers, glass wool and in glass fiber reinforced cement (Majumdar and Ryder,1968).

Boccaccini et al (1997) produced glass fibers from CRT panel and cone glasses. One possible use is application of these fibers to fabricate textiles for radiation protection. Scarinci et al (2000) obtained glass fibers by vitrification of MSW ashes and sludges plus glass cullet. Cioffi et al.(2001) produced glass fibers from industrial waste. Hreglich et al.(2001) proposed using panel glass from cathode ray tubes (CRT) to produce A-type glass fibers, using glass cullet. Majumdar et al.(1977) reported that superior performance of zirconosilicate glass fibers as reinforcement for cements is linked to the inherent alkali-resistance of the glass. Marabini et al (1998) produced glass wool by melting talc and chromite mine tailings at 1350C.

Glass foams are obtained by adding suitable blowing agents to glass powder, which is then heated above the softening point to release the gas and achieve the porous structure (Spiridonov and Orlova, 2003). A serious challenge when using glass derived from wastes is the tendency of these glasses to crystallize upon heating. If crystallization occurs during foaming, homogeneity and reproducibility problems arise, thus cullet has been the most accepted starting material. Ducman and Kovacevic (1997) manufactured glass foams by adding calcium carbonate or MnO_2 foaming agents to powdered recycled soda-lime glass. The processing temperature depended upon the foaming agent (750C – 780C) for $CaCO_3$ and (850-880C) for the MnO_2.

In addition, several authors (*Loewenstein, 1973,* Forsberg et al., 1996, Sheng,1997, Jahagirdar and Wattal, 1998, Luo et al, 2001, Sheng, 2001, Sheng, 2002, Raj et al, 2006 and Kaushik, 2006) have investigated the potential for incorporating low to high level nuclear wastes in borate glasses, where waste loading, product chemical durability and melt viscosity were considered the developmental and application concerns. Vaswani et al (1979) and Samuel et al (1988) concluded that a molar glass composition of 43.5 SiO_2, 17.5 NaO_2, and 7.0 B_2O_3 with added MnO (10%) and TiO_2 (6.0%) was ideal for immobilization of some nuclear waste compositions. Mellinger and Chick (1979) reported that addition of transition element oxides (e.g. MnO_2, TiO_2) to borate glasses reduces melt viscosity, while TiO_2 can also act as a network former. Shah et al. (1995) reported producing amorphous glasses from reagent grade mixtures of MnO_2 + Na_2CO_3 + H_3BO_3 + SiO_2 using less than 20 mol.% of MnO_2, while a crystalline phase resulted from mixtures with > 20mol.% MnO_2 in the glass. Boccaccini et al (2004) investigated production of borosilicate and lead silicate glass-matrix composites for nuclear waste encapsulation.

The main aim of this research is to study the potential for producing a borosilicate glass from manganese crust tailings by employing the vitrification method and to determine representative physical and chemical properties of the product that may indicate potential commercial applications. In this study, raw materials chemistry, glass chemical durability and strength and potential product applications were the main factors considered during glass formulation and end-product design.

2. Materials and Methods

Considering the economics of glass production, the objective was to produce acceptable quality glass, while employing as high a percentage of tailings as possible. In addition, as little boric acid flux was employed as was feasible. The raw crust tailings were melted in a lab furnace at approximately 1285C degrees. Laboratory vitrification experiments were conducted using various proportions of crust tailings (slimes), silica (fine sand) and the additive flux (boric acid, $B_2O_3 . 3H_2O$). Upon reaching the melting temperature, the melts were held for 1h, then permitted to air cool to 30C degrees above (An et al, 2006) the glass transition temperature (T_g)(Dwivedi et al, 1994) of 540C, as determined from the DTA curve, where they were annealed for 1h, then air cooled to room temperature. The weight of the 5-15 glass mixture was determined before firing and compared to the weight after firing at 1160C.

Glass was cut from the centers of the samples and polished for further experiments. Glass sample 5-15 was tested using a DSC/TGA SDT Q600 series thermal gravimetric analyzer (TGA) to determine weight loss during the melting process as a function of temperature. The crystalline phase composition of a powdered (< 100 microns) 5-15 sample was investigated using a X-ray powder diffraction (Phillips PW 1720 generator unit). The DSC/TGA SDT Q600 series thermal gravimetric analyzer was also employed to determine the thermal properties of glass 5-15. Curves were recorded in an Argon atmosphere, using a powder (<200um) sample, at a heating rate of 10C/min.and with Al_2O_3 oxide as the reference material.

Average density of ten samples of glass 5-15 was determined by the Archimedes method (ASTM, 1992). The Vickers hardness value (H_v) was determined in (GPa) according to the equation (Rincon and Capel, 1985). The indentation from the Vickers indenter permits a quick and reliable method of determining fracture toughness (K_c) as the diamond edges create cracks that are easily visible on polished surfaces. Fracture toughness was determined in (MPa $m^{1/2}$) using the equation by Evans and Charles (1976). The elastic modulus (E), a measure of strength, was determined theoretically in GPa for glass sample 5-15 by substituting values determined for H_v and K_c into four known equations by; Lankford (1984), Miyoshi, Sagawa and Sassa (1989), Marshall and Evans (1981) and Anstis, Chantilkul, Lawn and Marshall (1981), then comparing the results for accuracy. The TEC value was determined theoretically using the factor of Appen method (McLaughlin and Moore, 1985). Chemical resistance of 2g of sample 5-15 ranging in particle size from 0.3 to 0.5 mm was determined as the average amount of weight reduction from 5 trials after holding samples at 95C for 1 h in 60 cm^3 of 0.01 mol/l NaOH and 0.01 m/l HCL solutions, respectively. The samples were weighed before immersion and washed, then dried and weighed again after immersion (Leroy et al, 2001).

Glass fibers were produced by re-melting parent glass at 1180C in an electrically heated platinum bushing. The fibers were pulled from the melt and collected into bundles on a rotating drum. Descriptions of the fracture surface of a 5-15 glass sample fiber and particle sizing of the raw slimes were achieved by using a Hitachi 4700-S electron scanning microscope (SEM). A 5-15 glass fiber sample was mounted in non-conductive carbon paste and sputter coated (387A thickness) with gold and palladium using a Quorum technologies Polaron SC7620 Mini Sputter Coater prior to SEM observation.

3. RESULTS AND DISCUSSION

Manganese crust tailings samples came from the BHP Billiton mine and mineral processing facility at Groote Eyland in Northern Australia. The original ore was mined from an ancient submarine manganese deposit that is presently exposed on Groote Eyland. The chemical compositions of the as-received tailings and boric acid analyzed by x-ray fluorescence spectroscopy (XRF) are provided in Table 1.

Figure 1 displays an SEM of mainly angular as-received manganese crust tailings, which reveals grain sizes ranging from 10 to 150microns. Grain sizes varied between shipments of tailings from the sources, thus larger grains may range from 100 to 300+ microns. Using a Malvern particle size analyzer, the silica sand additive was found to be 100% passing 100microns, while the boric acid crystals were 100% passing 20 microns.

Thus, crushing of waste materials and additives was deemed unnecessary, which precluded an extra stage and crushing costs. In addition, considering that Potts (1944) suggested that homogeneous melts are most likely when grain sizing is relatively similar between all components of the melt mixtures, the likelihood of achieving a homogeneous melt was increased for the study melt. Several trials were conducted at various particle sizes and it appeared that between 100 and 200 microns produced the optimal quality glass product following criteria described by (Shelby, 2005).

Table 1. Chemical compositions of manganese crust tailings and boric acid flux

Oxide (wt.%)	Slimes	Boric acid	Oxide (wt.%)	Slimes	Boric acid
SiO2	28.1		K2O	0.34	
Al2O3	20.3		CaO	0.04	
MnO2	36.0		MgO	0.14	
Fe2O3	5.6		TiO2	0.48	
B2O3		76.0	H2O	8.36	24.0
Na2O			Others	0.04	
BaO	0.6				

The most common borosilicate glass is made by melting a mixture of SiO_2 (silica), Na_2CO_3 (soda) and lime (CaO) at over 821C to produce a glass with density of 2.23g/cm^3 (Shelby, 2005), while the commonly used SIMAX borosilicate glass composition is 80wt% SiO_2, 13 wt.% B_2O_3, 2.4 wt% Al_2O_3 and 4.3 wt% $Na_2O + K_2O$ (Gnizdo, 1956). The weight percentages for slimes are; SiO_2= 28.1%, Al_2O_3 = 20.3%, MnO_2 = 36%, Fe_2O_3 = 5.6%, H_2O = 8% and others = 2%. Thus, considering the low weight percentage of silica, the boric, alumina, iron and manganese oxides in the slimes and flux, plus silica additive must act collectively as major to minor glass formers and modifiers to enable glass production.

Numerous trials suggested that glass samples (5-15, 75/20/5 wt.% of tailings/flux/silica, 5-16, 80/20/0 wt.% and 5-17, 50/30/20 wt.%) exhibited an acceptable, bubble-free appearance, while glass 5-15 was chosen as representative for extensive testing. All glass products were a reddish-brown to black colour in solid form and light to dark brown in powder form. Density of the three study glasses ranged from 2.8 to 3.0 g/cm^3, which is similar to other known (Huang et al. 2004) waste glasses. Considering that volume reduction of wastes is often an important focus of waste treatment, the weight loss from initial to post melting was calculated as approximately 23%.

Figure 1. SEM micrograph of as-received manganese crust tailings.

Fe_2O_3 (eq.1)(Sglavo et al, 2000), MnO_2 (eq.2)(Shah et al, 1995) and boric acid (eq.3) release gases during the melting phase, which should aid in achieving melt homogenization in an industrial setting, although the frothing of the melt has caused loss of product to the sides of the melt container and negated the measurement of viscosity of the melt under laboratory conditions. Specifically, bubbles and frothing produced during melting were due to gaseous emissions created by similar activities to those reported as reduction of iron (Sglavo et al, 2000), reduction of manganese oxides (Shah et al, 1995), and breakdown of clay silicates (Appendino et al., 2004).

(1) $MnO_2 + H_2$ yields $MnO + H_2O$, $4MnO_2$ yields $2Mn_2O_3 + O_2$ (680 to 800C) (Handbook, 1997)
(2) $2Fe_2O_3 + \Delta$ yields $4Fe + 3O_2$ (Handbook, 1997)
(3) $B_2O_3 . 3H_2O + \Delta$ (>170C °) yields $B_2O_3 . H_2O + H_2O$ (Handbook, 1997)

According to (Sheng et al, 2003), viscosity is directly related to the amount of formers present in the melt, while the higher the Na_2O content, the lower the expected viscosity. Pouring of the study melt into molds at a 1160C degree melting temperature was observed as being relatively easy and acceptable for laboratory studies, thus the lack of sodium oxide content may have promoted the lower viscosity levels. It has been reported (Leroy et al, 2001) that addition of modifier oxides is useful to moderate the viscosity and facilitate melting in melts with >86wt.% $SiO_2 + Al_2O_3$. The study melts contained only 48wt% $SiO_2 + Al_2O_3$, thus could be considered of lower viscosity and already contained modifiers that would assist in maintaining a moderate viscosity to enhance melting efficiency and lower processing costs. In addition, study glass 5-15 contains potentially useful transition oxides TiO_2, and MnO_2 that can reduce viscosity and enhance melting, along with TiO_2 potentially acting as an additional network former in the production of glass compositions for possible nuclear waste immobilization.

When glass is cooled from an elevated temperature, internal stresses (thermal stresses) may be introduced resulting from the difference in cooling rate and thermal contraction between the surface and interior regions. These stresses may lead to product fracture (thermal shock). Cooling the product at a sufficiently slow rate may avoid occurrences of thermal stresses (Callister, 1985).

Kuraoka et al. (1998) has reported that sodium borosilicate glass as a phase separate glass is one of the most common source materials for porous glasses and that phase separation is the key step in preparing porous glasses. In addition, Chen et al (2006) reported that calcium oxide and alumina strongly affect phase separation, thus the noticeably alumina content of the slimes could play a role in determining phase separation in the study glasses. Following data reported by (Rawson, 1991), study glass 5-15, having been formed from a melt containing more than one significant glass former (SiO_2 and B_2O_3), is a likely candidate for phase separation. In addition, Tomozawa (1979) has suggested that micro-phase separation occurs even in quenched homogeneous soda-lime-silicate glasses, resulting in borate-rich and silica-rich phases with an interface phase between the two major phases. In response to this, Yazawa et al (1999) has described an approach to avoiding phase separation during cooling from the melt, as far as possible, by quenching with high cooling rates. Thus, if the study melt had been quenched, phase separation may have been less likely to occur. Contrasting this, a slow cooling rate was employed in this study as an attempt to induce phase separation and may have the added benefit of relieving unnecessary thermal stresses. In addition, slow cooling of the melt negates one processing step, which can decrease overall production costs.

The XRD image in Figure 2 indicates that with 75 wt% waste oxide loading, an amorphous material is produced as glass 5-15, with no significant crystallization, which is considered a relatively high waste loading (Romero and Rincon,1998, Shao et al, 2005) and makes the vitrification approach more appealing Figure 3 is an optical micrograph (20 times) of glass 5-15 showing the glass has apparently undergone amorphous phase separation (glass-in-glass phase separation).

The glass appears to contain larger grains inter-dispersed throughout the glass matrix. Figure 4a displays the EDX spectrum for glass sample 5-15, while Figure 4b shows the qualitative elemental percentages. The absence of the atomic element boron on the EDX is attributed to this elements being too light to be detected by the Hitachi S-4700 SEM/EDX apparatus. The Au and Pd values are due to the sputter coating. The DTA graph in Figure 5 suggests the glass transition temperature (T_g) to be at 540C for glass 5-15. The T_g signifies the temperature below which the glassy phase remains intact and the region where the glass will not devitrify under repository conditions, since the ambient temperature is less than the T_g. The dilatometric softening point (T_{ds}) is at 570C, and melting temperature is at approximately1120C. These DTA values are very similar to the zinc borosilicate glass identified by An et al. (2006).

Table 2 displays the values determined in the lab for density, Vickers micro hardness (H_v), indentation strength (K_c), elastic modulus (E) and thermal expansion coefficient (TEC), which indicates that the study glass is relatively similar in strength to known glasses. Considering the importance of durability in most glass applications, it was attempted to fabricate glasses with as high a level of durability as possible, considering the materials being used, equipment available and melting conditions applied. Study glass 5-15 exhibits a chemical durability quite similar to known glasses (Table 3) and thus is considered promising.

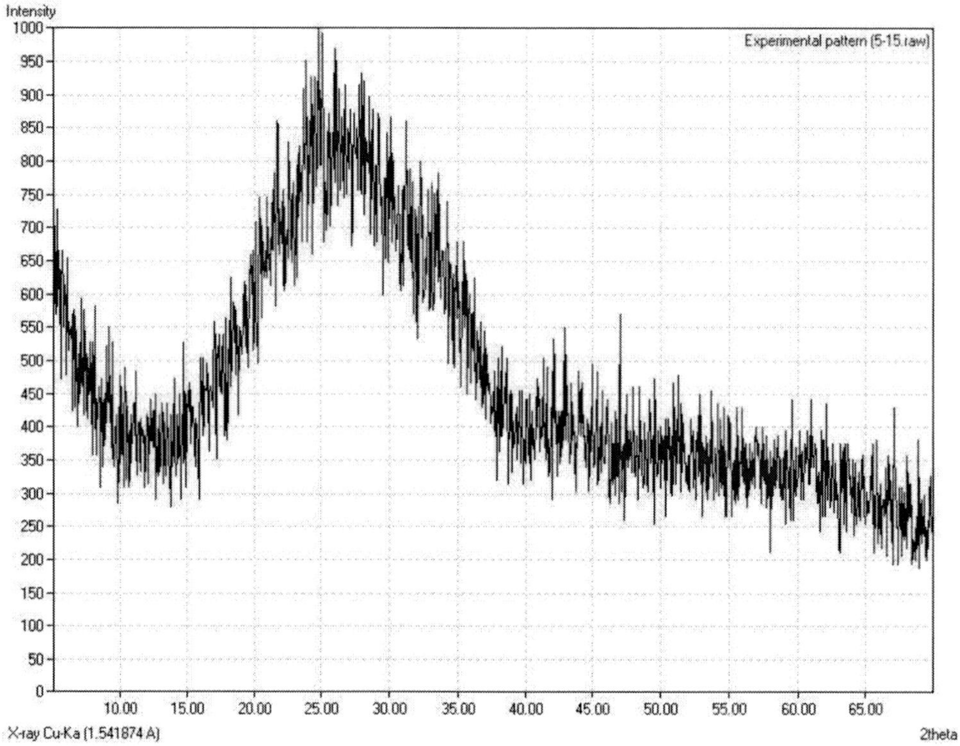

Figure 2. XRD pattern of glass 5-15.

Figure 3. Optical micrograph of study glass sample 5-15.

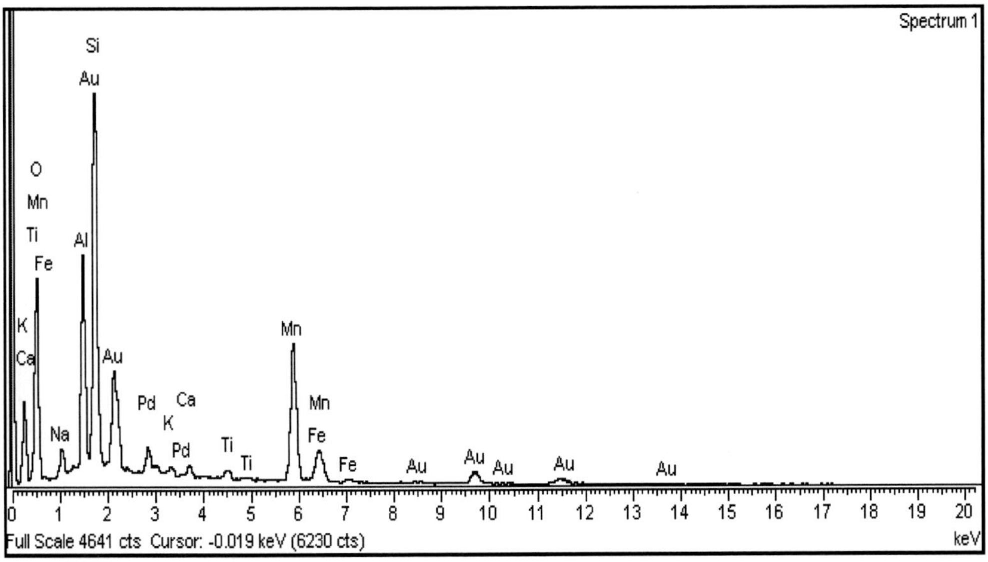

Figure 4a. EDX micrograph of study glass 5-15.

Element	Weight%	Atomic%
O K	54.45	71.90
Na K	2.63	2.42
Al K	8.94	7.00
Si K	14.83	11.16
K K	0.43	0.23
Ca K	0.56	0.30
Ti K	0.56	0.25
Mn K	14.81	5.69
Fe K	2.78	1.05
Totals	100.00	

Figure 4b. EDX elemental percentages.

Table 2. Physical properties of glass 5-15 compared to known glasses

Samples	Density g/cm3	Vickers Hardness GPa	Kc MPa.m1/2	E GPa	TEC x 10-6
5-15	2.3	4.8	0.66	92	6.89
A		6.2			8.42
B	2.9	4.8	0.9		9.5
C	3.41	6.85	2.02	171	

Sources: A (Scarinci, 2000), B (Park and Heo, 2002), C (Romero and Rincon (1998).

Table 3. Comparison of chemical durability of the study glass to commercial glasses based on weight-loss measurements

Samples mg/cm^2	NaOH 5% (6 h)	HCl 5% (24 h)	Na_2CO_3 0.2% N(6h)
Soda lime glasses*	0. 80	0. 05	1.50
Basalt glass*	1.10	0.90	0.10
5-15 glass	0 .76	0 .40	0.86

Source: Beal and Rittler (1976)*.

The relatively low TEC value of 3.89 x 10^{-6} for glass 5-15 could be related to the lack of sodium oxide content in the glass mixture. The observed physiochemical character of laboratory glass 5-15 may be at the lower end of potential strength and durability values, since production of homogeneous glass is restricted by the effects of volatile melts and the possible development of glass defects. Observed volatility of glass 5-15 was not appreciably lowered by melting at 1160C, since large volumes of oxygen are released from clay dehydration and transition of Fe_3O_4 to Fe_2O_3 at temperatures lower that 1160C. Following data by Gelhoff and Thomas (1926) and Morey (1954), the addition of boric acid flux as a glass component could increase the strength of the study glasses. In addition, the use of boric acid probably weakened the structure of the silica and lead to the lower melting point observed for the glass mixture. This feature translates to considerable energy savings during the glass production process.

Considering the findings of Ballard and Pye (1976), the manganese and iron oxide content of glass 5-15 could enhance the alkaline-resistance of the glass and increase its durability. In addition, following the findings of Paul and Youssefi (1978), the low initial wt% of CaO earth alkali oxide in the study melt and the addition of large volumes of FeO and MnO to the melt could lead to increased alkaline durability. The acceptable durability values of glass 5-15 could also be partially due to the effects of the alumina modifier on the melt as noted by Shelly (2005).

Marabini et al. (1998) produced glass wool by melting talc and chromite mine tailings at 1350C. In addition, the authors noted that rock wool was produced from synthesis with the main components being; CaO, SiO_2, Fe_2O_3, MgO and Al_2O_3 in such proportions as to ensure a characteristic basicity index given by the ratio; $SiO_2 + Al_2O_3/CaO + MgO + Fe_{total} = 1$). It was stated that this ratio must be achieved by addition of appropriate oxides. Following this approach, by including manganese oxide with total iron oxide content, the oxide ratio for slimes to produce glass wool is 1.6, relatively close to the desired ratio of 1.0, although if boric oxide were added as a glass former with silica, the ratio would move higher. This at least provides an indication that slimes may be a candidate for glass wool production. Considering that MnO_2 and Fe_2O_3 are components of the 5-15 glass mixture, the melt contains blowing agents, which create foaming between 850 and 960C, as evidenced by the endothermic peak in Figure 5. This evolution of gas could create sufficient foaming to enable production of a glass foam product. It remains to establish a satisfactory relation between process parameters and potential glass foam microstructure.

The feasibility of producing glass fibers from crust tailings to manufacture glass wool as an insulation product will depend upon the price received for the product, compared to the total costs of production. The total costs of production would consist of tailings processing costs to produce the glass fibers, costs to convert fibers into glass wool, cost of boric acid additive, plus waste and product transportation costs. The wastes are considered at no cost since they are deemed a liability to the processing company, while boric acid costs are as low as a few dollars in bulk tonnage purchases. In Figure 6, is reported a SEM micrograph of the fracture surface of a 5-15 borosilicate glass fiber, whose morphology is typical for a glass and the three regions of the mirror, mist and hackle are quite visible (Bradt and Tressler, 1994).

A primary objective of the present research has been to produce glassy materials with applications as porous glasses, thus a slow cooling approach was employed to promote phase separation.

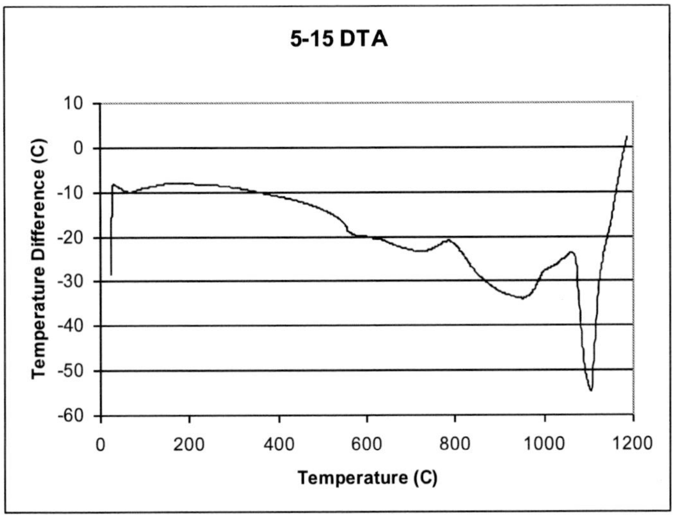

Figure 5. DTA graph of glass 5-15.

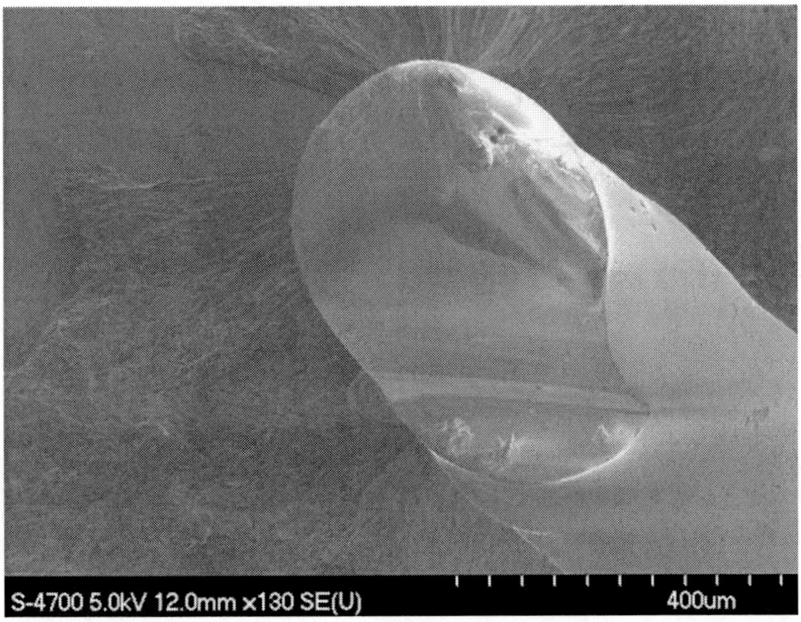

Figure 6. SEM of 5-15 glass fiber.

Contrasting this, in nuclear glass research, the glass components are chosen to avoid phase separation, due to possible deleterious effects of phase separation on glass durability (Cozzi and Jantzen,1998), where quenching is considered a useful approach to avoiding separation of the melt. In addition, a Statistical Process Control (SPC) system has been developed using process and product quality models to determine the performance of Defence Waste Processing Facility (DWPF) waste glasses, based upon composition (Jantzen and Brown, 1999). Also, Peeler and Hrma (1996) have provided background data on phase separation in multi-component borosilicate glasses, while Cozzi and Jantzen (1998) reported that DWPF glasses are formulated to utilize the ability of sludge components to produce homogeneous glasses with frits. They described adding a phase-separated (Frit 202) to the DWPF, resulting in two possible homogeneous glasses. In addition, Vienna et al. (1996) reported that a small alumina addition to some homogeneous glasses could significantly increase glass durability. Respecting the present research, the slimes-based waste with a boric oxide flux contains several components (e.g. alumina), which suggests that may be fashioned into a frit similar to Frit 202.

Conclusion and Recommendations

Manganese crust tailings can be vitrified into a borosilicate glass material which exhibits physiochemical properties that suggest marketable products may be produced by recycling this waste product. The expense of the flux material may be partially offset by the potential sale of products produced from the study glass. Production of the study glass reduced the tailings volume by over 23 percent.

Employing boric acid as a flux decreased the theoretical melting temperature by over 100C degree and enabled glass production. The glasses displayed acceptable durability and workability when 20 to 30wt% boric acid was added to the mixture.

Future research is required to determine if glass 5-15 can be fashioned into glass foam and glass wool products. Additional research is also required to refine the fiber production process using the 5-15 glass melt.

Future research should involve using NMR, TEM and Raman spectroscopy to identify the roles played by various oxides in developing glass structure as a basis for determining whether the study waste-based glass can be employed to consolidate radioactive waste materials. In addition, the suitability of slimes for development of a glass frit for application in nuclear waste management should be examined.

Acknowledgment

Appreciation is extended for numerous measurements being performed using equipment made available by Dr. Jeff Dahn of the Department of Physics and Atmospheric Sciences at Dalhousie University. A note of thanks is also given to Dr. Rebecca Jamieson of the Earth Sciences Department at Dalhousie for use of the optical microscope.

REFERENCES

An, J. Park, J. Kim, J., Hong, K.S. (2006). Effects of Bi_2O_3 and Na_2O on the thermal and dielectric properties of zinc borosilicate glass for plasma display panels, *J. Am. Cer. Soc.*, vol. 89, no.15, pp.3658-3661.

Appendino,P., Ferraris, M., Matekovits, I., Salvo, M.(2004). Production of glass-ceramic bodies from the bottom ashes of municipal waste incinerators, *J. of the European Ceramic Soc.*, vol.24, pp.803-810.

ASTM C693 (1992). Standard test method for density of glass by buoyancy.

Ballard, C.P., Pye, L.D. (1976). Glass formation in iron-soda-silica system, *J. Amer. Cer. Soc.,* vol.59, pp.266-267.

Beal, G.H., Rittler, H.L. (1976). Basalt glass-ceramics. *American Ceramic Society Bulletin*, vol.55, no.6, pp.579-582.

Besmann, T.M., Spear, K.E.(2002). Thermochemical modeling of oxide glasses, *J. Am. Cer.* Soc., vol.85, pp.2887-2894.

Blight, G.E., Fourie, A.B. (2002). A review of catastrophic flow failures of deposits of mine waste and municipal refuse, U. of Witswatersand, Johannesburg, S.A.

Boccaccini, A.R., Bucker, M., Bossert, J., Marszalek, K. (1997). Glass matrix composites from coal fly ash and waste glass, *Waste Mgmt.* (Oxford), 17(1), pp.39-45.

Boccaccini, A.R., Bernardo, E., Blain, L. (2004). Borosilicate and lead silicate glass matrix composites containing pyrochlore phases for nuclear waste encapsulation, *J. of Nuclear Mat.,* vol.327, no.2-3, May, pp.148-158.

Bradt, R.C., Tressler, R.E. (1994). Fractography of Glass, Plenum Press, New York, p.157.

Callister, W.D. Jr. (1985). *Materials Science and Engineering: An Introduction*. John Wiley and Sons, Inc.

Cheng, T.W. (2003). Combined glassification of EAF dust and incinerator fly ash, *Chemosphere*, vol.50, pp.47-51.

Chen, D., Masui, H., Miyoshi, H., Akai, T., Yazawa, T. (2006). Extraction of heavy metal ions from waste coloured glass through phase separation, *Waste Management*, Jan., vol. 26, no.9, pp.1017-1023.

Cozzi, A.D., Jantzen, C.M. (1998). *Glass durability and the compositional continuum between a phase separated frit and a homogeneous glass*, Westinghouse Savannah River Company, Aiken, S.C., 29808, WSRC-MS-98-00836, Rev.1.

Cunnane, J.C., Bates, J.K., Ebert, W.L. (1992). H*igh-level nuclear waste borosilicate glass: A compendium of characteristics,* ANL/CMT/CP-76415, Argonne National Laboratory, Argonne, IL.

Dwivedi, A., Berta, Y., Speyer, R.F. (1994). Effect of controlled crystallization on the chemical durability of a lead-containing waste glass, *J. of Mat. Sci.*, vol.29, pp.2304-2308.

Evans, A.G., Charles, E.A. (1976). Fracture toughness determination by indentation, *Journal of the American Ceramic Society*, vol.59, no.7, pp. 371-372.

Ellison, A.J.G., Mazer, J.J., Ebert, W.L. (1994). Effect of glass composition on waste form durability, *a critical review*, ANL-94/28.

Feng, X., Cunnane, J.C., Bates, J.K. (1993). A literature review of surface alteration layer effects on waste glass behavior, *American Cer. Soc.* 1993 Annual Meeting, Cincinnati, Ohio, USA.

Forsberg, C.W., Beahm, E.C., Parker, G.W., Elam, K.R. (1996). Conversion of radioactive and hazardous wastes into borosilicate glass using the glass oxidation and dissolution system, *Waste Mgmt.*, vol.16, no.7, pp.615-623.

Francis, A.A. (2004). Crystallization kinetic of glass particles from a mixture of coal ash and soda-lime cullet glass, *Journal of non-crystalline solids*, vol.333, Feb.1, pp.187-193.

Frugifer, P., Gordan, N., Vernaz, E., Larch, F. (2002). Influence of compositional variations on the initial alteration rate of vitrified domestic waste incinerated fly ash, *Waste Mgmt.* vol. 22, pp.137-142.

Gelhoff, G., Thomas, M. (1926). Physical properties of glasses related to their composition, *Z.Tech. Fiz.* vol.7, pp.105-126.

Gnizdo, M. (1956). Simax — A new borosilicate glass of the highest quality, *Glass and Ceramics*, vol.13, no.12, Dec., pp.539-540.

Hood, H.P., Norberg, M.E. (1984). U.S. Patent No. 2,106,744, U.S. Patent No. 2,286,275.

Huang, W., Day, D.E., Ray, C.S., Kim, C.W., Mogus-Milankovic, A.M. (2004). Vitrification of high chrome oxide nuclear waste in iron phosphate glasses, *J. Nucl. Mat.*, vol 327, pp. 46-57.

Jahagirdar, P.B., Wattal, P.K. (1998). Vitrificaiton of sulphate bearing high level wastes in borosilicate matrix, *Waste Management*, vol.18, pp.265-273.

Kaushik, C.P., Mishra, R.K., Sengupta, P., Kumar, A., Das, D., Kate, G.B., Raj, K. (2006). Barium borosilicate glass: A potential matrix for immobilization of sulfate bearing high-level radioactive liquid waste, *J. of Nuclear Materials*, vol. 358, nos.2-3, pp.129-138.

Kuraoka, K., Qun, Z., Kushibe, K., Yazawa, T. (1998). *Separation Science and Technology*, vol.33, pp.297.

Jantzen, J.B., Brown, K.G.(1999). Impact of phase separation on waste glass durability, Westinghouse Savannah River Company, Aiken, SC 29808, WSRC-MS-99-00333, Contract No. DE-AC09-96SR18500 with the U. S. Department of Energy.

Leroy, C., Ferro, M.C., Monteiro, R.C.C., Fernades, M.H.V. (2001). Production of glass-ceramics from coal fly ashes, *J. of the European Ceramic Society*, vol.21, no.2, pp.195-202.

Loewenstein, K.L. (1973). The Manufacturing Technology of Continuous Glass Fibers. New York: Elsevier Scientific, pp. 2-94.

Luckscheiter, B., Nesovic, M. (1996). Development of glasses for the vitrification of high level liquid waste (HLLW) in a joule heated ceramic melter, *Waste Mgmt.*, vol.16, no.7, pp.571-578.

Luo, S., Sheng, J., Tang, B. (2001). A comparison of HLW-glass and PWR-borate waste glass, *J. of Nuclear Mats.*, vol.298, pp. 180-183.

McLaughlin, P.O., Moore, D.T. (1985). Models for the thermal expansion coefficient and temperature coefficient of the refractive index in gradient-index glass, *Applied Optics,* vol.24, no.24, pp.4342-4346.

Majumdar, A.J., Ryder, J.F. (1968). Glass Fiber Reinforcement of Cement Products, BRS current paper 67/68, *Sept. 1968 and Glass Technology*, vol. 3, no. 3, June, p.78-89.

Majumdar, A.J., West, J.M., Larner, L.J. (1977). Properties of glass fibres in cement environment, *Journal of Materials Science*, vol.12, no.5, May, pp.927-936.

Marabini, A.M., Plescia, P., Maccari, D., Burragato, F., Pelino, M. (1998). New materials from industrial mining wastes: glass-ceramic and glass and rock-wool fiber, *Intl. J. of Min. Process.*, vol.53: 1-2, Feb., pp.121-134.

Mellinger, G.B., Chick, I.A. (1979). In: *Ceramics in Nuclear Waste Management*, CONF-790420 Cincinnati,, T.D. Chilalia and J.E. Mendel (eds.), Technical Information Center, Cincinnati, Ohio, p.213.

Mogus-Milankovic, A., Rajic, M., Drasner, A., Trojko, R., Day, D.E. (1998). Crystallization of iron phosphate glasses, *Phys. Chem. Glasses*, vol. 39 (2), pp.70-75.

Mohr, J.G., Rowe, W.P. (1978). Fiberglass. Atlanta: Van Nostrand Reindhold, p.13.

Morey, G.W.(1954). *Properties of Glass*, Reinhold, New York, pp.304-344.

Park, Y.J., Heo, J.(2002). Nucleation and crystallization kinetics of glass from incinerator fly ash waste, Sch. Of Environmental Sci.. and Engineering, Pohang Univ. of Science and Technol., *Ceramics Intl.*, vol.28, no.6, pp.669-673.

Paul, A., Youssefi, A. (1978). Alkaline durability of some silicate glasses containing CaO, FeO and MnO, *J. Materials Sci.*, vol.13, pp.97- 107.

Peeler, D.K., Hrma, P.R. (1996). Compositional range of durable borosilicate simulated waste glasses, In: *Emerging Technologies in Hazardous Waste Management VI*, eds. D.W.Tedder and F.G. Pohland, pp.323-338, American Academy of Environmental Engineers.

Pelino, M., Cantalini, C., Abbruzzese, C., Plescia, P. (1996). Treatment and recycling of goethite waste arising from the hydrometallurgy of zinc, *Hydrometallurgy*, vol.40, pp.25-31.

Handbook (1997). *Perry's Chemical Engineer's Handbook*, 7th Ed., Edited by R.H. Perry and D.W. Green, McGraw-Hill.

Potts, J.C., Brookover, G., Burch, O.G.(1944). Melting rate of soda lime glasses as influenced by grain sizes of raw materials and addition of cullet, *J. Am. Cer. Soc.* vol. 27, pp.225-231.

Raigon-Pichardo, M.,Garcia-Ramos, G., Sanchez-Soto, P.J. (1996). Characterization of a waste washing solid product of mining granitic tin-bearing sands and its application as a ceramic raw material, *Resources Conservation and Recycling*, vol.17, pp.109-124.

Raj, K., Prasad, K.K., Bansal, N.K. (2006). Radioactive waste management practices in India, *Nuc. Engin. and Design*, vol.236, pp.914-930.

Rawson, H. (1991). Glasses and Their Applications, *The Institute of Metals*, 159p.

Riba, I., Del Valls, T.A., Forja, J.M., Gomez-Parra, A. (2002). Influence of the Aznalcóllar mining spill on the vertical distribution of heavy metals in sediments from the Guadalquivir estuary (SW Spain), *Marine Pollution Bulletin*, vol. 44, no.1, pp. 39-47.

Rincon, J. Ma., Capel, F. (1985). Micro-indentation behavior K_{ic} factor determination and microstructure analyses of some LiO_2-SiO_2 glass-ceramic materials, *Cer. Intl.*, vol.11, no.3, pp.97-102.

Romero, M., Rincon, J.M. (1998). Preparation and properties of high iron oxide content glasses obtained from industrial wastes, *Eur. Cer. Soc.*, 18(12), pp.153-160.

Samuel, M.T., Rana, D.S., Raj, K., Yeotiker, R.G. (1988). Performance of solidification of high level waste forms and engineered barriers under repository conditions. In: *IAEA 2nd Research Coordinated Meeting,* Australian Nuclear Research Institute, Sydney, Australia.

Scarinci, G., Brusatin, G., Barbieri, L., Corradi, A., Lancellotti, I., Colombo, P., Hreglich, S., Dall'Igna, R. (2000). Vitrification of industrial and natural wastes with production of glass fibers, *J. Euro. Cer. Soc.*, vol. 20, nos. (14-15), pp.2485-2490.

Sglavo,V.M., Campostrinin, R., Maurina, S., Carturan,G., Monagheddu,M., Budroni,G., Cocco, G. (2000). Bauxite "red mud" in the ceramic industry, Part I: thermal behavior, *J. of the European Ceramic Society*, 20, no.3, pp.235-244.

Shah, J.G., Patki, V.A., Raj, K. (1995). DTA, powder XRD and SEM study of manganese-containing borate glasses, *Waste Management*, vol.15, nos. 5/6, pp.417- 421.

Shao, H., Liang, K., Peng, F., Zhou, A., Hu, A. (2005). Production and properties of cordierite-based glass-ceramics from gold tailings, *Minerals Engineering*, vol.18, pp.635-637.

Shelby, J.E. (2005). *Introduction to Glass Science and Technology*, Royal Society of Chemistry, London, England, 270 p.

Sheng, J. (1997). Vitrification of borate waste from PWR nuclear power plant, PhD Thesis, China Institute of Atomic Energy, Beijing, China.

Sheng, J. (2001). Vitrification of borate waste from a nuclear power plant using coal fly ash (I) glass formulation development, *Fuel*, vol.80, no.10, Aug., pp.1365-1369.

Sheng, J. (2002). Vitrification of borate waste from nuclear power plant using fly ash, (II), leaching behavior of the FA30 glass, *Fuel,* vol.81, pp.253-256.

Sheng, J., Huang, B., Zhang, J., Zhang, H., Sheng, J., Yu, S., Zhang, M. (2003). Production of glass from coal fly ash, *Fuel*, vol.82, no.2, Jan., pp.181-185.

Simon, M., Martin, F., Ortiz, I., Garcia, I., Fernandez, J., Fernandez, E., Dorronsoro, C., Aguilar, J. (2001). Soil pollution by oxidation of tailings from toxic spill of a pyrite mine, *The Science of the Total Environment*, vol. 279, nos.(1-3), pp. 63-74.

Spiridonov, Y.A., Orlova, L.A. (2003). Problems of foam glass production, *Glass and Ceramics*, vol.60, nos. 9-10, Oct., pp.10-11.

Tomozawa, M. (1979). Phase separation in glasses, In: *Treatise on Materials Science and Technology*, vol.17, Glass II, (eds.) M.Tomozawa and R.H.Doremus, Academic, New York, p.71.

Vaswani, G.A., Jahgirdar, P.B., Rastogi, R.C., Sunder-Rajan, N.S. (1979). Development of suitable vitrified radioactive waste products with low formation temperatures and improved leach resistance – A practical approach, BARC Report 1028.

Yazawa, T., Kuraoka, K., Du, W.D. (1999). Effect of cooling rate on pore distribution in quenched sodium borosilicate glasses, *J. Phys. Chem.* B103, p.9841-9845.

In: Characterization and Development of Novel Materials
Editor: Rafiq Islam
ISBN: 978-1-61324-149-3

Chapter 11

Isolation and Characterization of Effective Bacteria for Solid Waste Degradation for Organic Manure

***M. Mizanur Rahman*[1*], *M. Mashiar Rahman*[1], *S. M. Y. Arafat*[2], *M. Z. H. Khan*[2], *Atiqur Rahman*[3] *and M. Safiur Rahman*[4]**

[1]Department of Biotechnology and Genetic Engineering, Islamic University, Kushtia, Bangladesh
[2]Department of Applied Chemistry and Chemical Technology, Islamic University, Kushtia, Bangladesh
[3]Department of Biotechnology, College of Engineering, Daegu, University, Kyungsan, Kyungbuk, Republic of Korea
[4]Environmental Analytical Chemistry Lab, NRCD, Institute of Nuclear Science and Technology, Bangladesh Atomic Energy Commission, Dhaka, Bangladesh

Abstract

The present investigation was conducted to find out the utilization of effective bacteria for solid kitchen waste degradation, and subsequent uses of degraded substances as organic manure or compost. Ten garbage samples were collected from different potential habitat of Kushtia and Jhenaidah regions of Bangladesh for isolating the effective garbage degrading bacteria. Three bacterial sample strains named as NAIUL, BCDump and BCSS (2) were isolated and cultured on Nutrient Agar and Czapek-Dox-Agar media. Optimum cultural conditions (pH of preferred medium, temperature, incubation period) of the bacteria were studied before massive multiplication. Characterizations of the strains were also studied by visual observation of colony, microscopic observation and biochemical tests identified the specific bacteria namely *Xanthomonas* spp., *Bacillus* sp., and *Pseudomonas* sp. from NAIUL, BCDump and BSS (2) strains, respectively. Gram staining reaction and motility test were studied by

* Corresponding author: Department of Environmental Engineering, Faculty of Engineering, Dalhousie University, D215-1360 Barrington St., Halifax, N.S. B3J 1Z1, Canada. Email: Safiur.Rahman@dal.ca.

microscopic observation. All the strains were Gram positive and motile. Biochemical tests ensured their physiological characteristics. To select the most effective strain for decomposition of waste, autoclave sterilized kitchen garbage was used and inoculated with bacterial suspension and culture pellet. Changes of color, odor, weight loss, volume loss, temperature and pH of decomposing garbage were noted. The highest decomposition activity in term of weight loss and volume loss was shown by suspension of BCDump sample strain. To observe the effect of various additives on decomposition, various concentrations (5 %, 10 % and 15 %) of sucrose and molasses solution were used along with bacterial suspension. Among different sucrose treatments, BCDump sample strain along with 15 % sucrose solution was found to be the most effective. As the sucrose solution treatment showed better performance as additive, various concentrations of molasses solution (5 %, 10 % and 15 %) were also tested along with bacterial suspension as an alternative and cheap source of carbohydrates. It was observed that 15% molasses treatment showed the best performance over the other treatments using BCDump sample strain. Decomposed garbage was used as compost to observe their effects on biomass production (fresh weight basis) of potatoes (*Solanum tuberosum*). Garbage decomposed by all strains was used as organic manure. From the present investigation, it can be concluded that useful bacteria might be isolated from the surrounding environment for friendly bioconversion of solid organic waste. To protect human health and natural environment from waste problems, the developed method of decomposition of waste as shown in this study is found to be very effective and useful.

Keywords: Bacteria, characterization, soil waste, organic compost

1. Introduction

The word "waste" refers to useless, unwanted or discarded materials which are no longer considered of sufficient value and are thrown away by the possessor (SDNP, 2005). Wastes include solids, liquids and gases, among them solid or semi solid forms are called "solid waste". Everyday, a huge quantity of waste generate in all the developed and developing countries (Paimela-Wheler, A 2004). In rapidly growing cities of the developing world, urban solid waste management is currently regarded as one of the most immediate problem. Various types of waste are causing adverse effect on living organisms and environment. As a result, human and animal diseases are occurring, the air and soil environment are spoiling and the entire natural ecosystem balance is disturbing. Previous studies showed that on an average, each person in urban areas produces half a kilogram of garbage each day (BBS, 1997; NILG and LGED, 2002). Only 10 % of this MSW (Municipal Solid Waste) is being recycled, while 80% goes into landfills, and 10 % is incinerated (US EPA, 1989). In low-income countries, the solid waste generation rate is average only 0.4 to 0.6 kg/person/day, as opposed to 0.7 to 1.8 kg/person/day in fully industrialized countries (Contreau, 1982; Arlosoroff, 1982; Blight and Mbande, 1996). Every day huge quantities of waste materials are generated in all cities and municipal areas of Bangladesh. In urban areas, solid waste has a very high organic content that ranged from 70- 85 %. The solid waste generation in urban areas in Bangladesh is growing with the growth of population as well as per capita GNP (Zurbrugg, 2002). Only Dhaka City Corporation estimates around 3000 tons of garbage per day from its 10.07 million populations having an area of 325 square km. Per capita garbage generation per day varies from 0.35 kg to 0.4 kg in Dhaka (NILG and LGED, 2002). The total generated waste is collected efficiency by formal system. About 10-15 % is recycled by formal system and 35-50 % is self disposable or illegal dumped (uncollected) (SDNP, 2005). Thus, under proper conditions this biodegradable fraction could be composed or co-composted for beneficial use as soil conditioners and biofertilizers (Parr and Hornick, 1992).

The microbial population of soils is made up of five major groups including bacteria, actinomycetes, fungi, algae and protozoa, and among these groups, bacteria are the most abundant group (Alexander, 1961). Among these microbes, bacteria are the most important microbes for decomposing waste. Bacteria use wastes for their own metabolism and finally they produce some simple and useful compounds from them which are important for soil health, plant growing and over all to keep well balance of natural ecosystem. Composting is the controlled biodegradation or transformation of organic material, usually under aerobic conditions by which a material is transformed into an end product which is stable and soil like material called compost.

Number of microbes along with rodents and insects play a vital role for solid waste degradation. Among them, bacteria play the most important role and therefore, the effective bacteria can be employed for planned decomposition of solid organic waste. Considering tremendous importance of garbage decomposition, this investigation has under taken in order to develop a reproducible protocol and to search for more active decomposer bacteria, which might be successfully and efficiently degrade the organic wastes and produce useful components for plant nutrition.

2. Materials and Methods

2.1. Sample Collection

Ten samples (soil/humus) were collected during 26 May, 2005 from ten different areas of Kushtia and Jhenaidah regions of Bangladesh. Each sample 100 g was collected in separate sterile screw cap tube and stored at 4°C. Each sample was marked according to their source and location as IUL (Garbage, In front of library, Islamic University Campus, Bangladesh), Dump (Garbage, behind of Begum Khaleda Zia Hall, Islamic University Campus , Kushtia), SS(2) (Soil surface, pond side of Bangabandhu Sheikh Mujib hall of Islamic University Campus, Bangladesh), NPTQ (Wet soil, drain side North para Teachers Quarter's, Islamic University Campus, Bangladesh), DR (Drain Side Garbage, Begum Fazilatunnessa hall, Islamic University Campus, Bangladesh), SS (Soil surface, behind the Science Faculty of Islamic University Campus, Bangladesh), DRS (Wet soil of drain, right side of Commerce Faculty of Islamic University Campus, Bangladesh), ZD (Wet soil of drain, behind of Ziaur Rahman Hall of Islamic university campus, Bangladesh), SG (Garbage, Sheikh Para, Jhenaidah, Bangladesh), and SLA (Refuse, slaughter house, Sheikhpara, Bangladesh). Of these samples, three bacterial samples such as IUL, Dump and SS(2) were showed better growth on their favorable medium; IUL showed on Nutrient Agar medium (NA), Dump showed on Basic Czapek-Dox-Agar medium (BCDA) and BCSS(2) showed on Basic Czapek-Dox-Agar medium (BCDA) and therefore, these three samples were marked as NAIUL, BCDump and BCSS(2), respectively.

2.2. Culture Method

Sterile dilution blanks were marked sequentially from 10^{-2} to 10^{-6}. Soil 1 g was added to the 10^{-2} dilution blank and shaked vigorously for at least 1 minute. Then the dilute was sedimented for a short period. 1 ml from this dilution was transferred to the 10^{-3} tube. Using a fresh sterile pipette, 1ml from the 10^{-3} dilution was transferred to the 10^{-4} dilution blank for each succeeding step, then from the 10^{-4} to the 10^{-5}, then from the 10^{-5} to the 10^{-6}. From each dilution tube, 1 ml of dilution fluid was transferred into different culture medium (NA, BCDA and PDA) and incubated at 37°C for 24 h. After successful growth of microorganisms, the colony were counted using a colony counter (Yc-2A, Prma optical works Ltd, Japan) and colony forming unit (g or ml) and the results per dilution count were recorded. All the plates were incubated at room temperature until next period (2 days) for successful growth observation. Then the viable titer was calculated and made a circle on the back of the each plate and assigned a number to each. Then each colony morphology e.g., size, shape, margin, elevation, consistency, color, transparency was determined. Each colony was performed to Gram stain to observe the cellular morphology and Gram reaction of the bacteria.

2.3. Pure Culture

One single colony was identified and re-streaked these cells as a primary inoculant's on the surface of a nutrient agar plate and basic czapek-dox agar (BCDA) media. Then the plates were incubated at 30°C or room temperature. Nutrient agar plates and basic czapek-dox agar (BCDA) re-streaked were examined for a pure culture. Each Gram stain determined which had the identical cell morphology and gram reaction as the original colony. At this point, another nutrient agar plate and basic czapek-dox agar (BCDA) were re-streaked with the correct colony, and then this step was repeated one above. After achieving a pure culture, the same colony was streak onto a nutrient agar and BCDA slant. These cultures were incubated for 1 day in the refrigerator. The isolates from soil were used for future experiments.

2.4. Biochemical tests

2.4.1. Catalase test

To detect soil bacteria, 2-3 ml of the hydrogen peroxide solution was poured in to a test tube. Using a sterile wooden stick, several colonies of the test organisms were removed from nutrient agar plate and basic czapek-dox agar (BCDA) medium and immersed in the hydrogen peroxide solution and bubbles formation was observed.

2.4.2. Lactose and Mannitol Fermentation Test

This test is used to differentiate the microorganisms that ferment carbohydrate (such as lactose, mannitol).

2.4.3. Voges Proskauer test

To detect the production of acetylmethylcarbinol (actoin), a natural product formed from pyruvic acid in the course of glucose fermentation. The buffered glucose broth with the organism was inoculated and incubated at 37°C for 3 days. Approximately 3 ml of alpha naphthol was added followed by 1 ml of 40 % KOH and mixed well for 30 minutes for results pink VP (+) and no change VP (-).

2.4.4. Urease test

To detect soil bacteria, a dense "milky" suspension of the test organism was prepared in 0.25 ml physiological saline in a small tube. A urease tablet was added into the tube and incubated at 35-37 °C for up to 4 h or overnight. Color change of the test organism was observed.

2.4.5. Indole test

Indole test was done for detecting the soil bacteria. The test organisms were cultured in tryptophan containing medium in a bijou bottle with 3 ml of sterile tryptone water. Kovac's reagent (4p-dimethylamino-benzaldehyde) 0.5 ml was added in the medium with gentle shaking and the color was observed.

2.4.6. Citrate utilization test

Citrate utilization test was based on the ability of an organism to use citrate as its only carbon source and ammonia as its only source of nitrogen. The isolated strain NAIUL showed positive citrate utilization test because media turned green to blue. BCSS(2) and BCDump did not turn the media green to blue and so they showed negative test.

3. Results and Discussion

Present studies were undertaken for the isolation and characterization of bacterial strains from different places of Kushtia and Jhenaidah regions. Bacterial growth depends upon various physiochemical conditions such as media, pH, temperature, incubation period, carbon source, etc. (Tchan, Y.T.2002)

Bacteria grow in a wide range of moisture level. In this study, it was found that the range of moisture content of collected samples was varied from 25.09 to 78.19 %. Maximum moisture content 78.19 % was obtained for NAIUL and BCDump samples, while minimum moisture content 25.09 % was noted for BCSS (2) sample.

Bacterial population of various soils is closely correlated with their moisture content. The maximum bacterial density is found in regions of fairly high moisture content and the optimum level for the actives of aerobic bacteria often is a 50 to 75 % of the soils moisture holding capacity (Alexander, 1961). Numbers of the genera *Pseudomonas*, *Achromobacter* and *Bacillus* are found in most aerobic soils; where conditions are anaerobic and moist *Clostridium* will occur. *Actinomycetes* showed a similar quantitative increase under such conditions (Berkeley and Cambelt, 1972).

In the present investigation, the growth of isolated strains was observed in various growth media like potato dextrose agar (PDA), nutrient agar (NA) and Czapek-Dox- agar (both acidic and basic). It was observed that the basic Czapek-Dox-agar (BCDA) was suitable for massive growth of BCSS (2) and BCDump strains (Table 1) and nutrient agar (NA) medium was suitable for the massive growth of NAIUL strain (Table 2).

The pH is a key factor for growing the bacteria in artificial media. In this study, the pH of selected two media (BCDA and NA medium) was optimized for culturing of the strains. The pH range of two medium was found to be 7-8. NA and BCDA medium at pH 7.2 and 7.6, respectively were found to be suitable for the maximum growth of bacterial strains. From the results, it was found that the pH of the samples of BCDump, BCSS(2) and NAIUL strain were 7.79, 7.95 and 7.86, respectively and possibly for this reason, these strains were also found to grow well in *in vitro* condition at pH 7-8 in BCDA and NA medium.

Bacteria can tolerate a soil reaction between pH levels 4 and 10, but the most favorable pH for the majority is just an alkaline side of neutrality. Bacteria such as *Thiobacillus thiooxidans* and *Acetobacter* sp. are capable of growing at the very low pH values of between pH level 0 and pH level 2 and some *Bacillus* sp. can grow at pH 11 (Berkeley and Cambelt, 1972).

Thermoactinomycetes grows only at temperatures between 50°C and 65°C and optimal growth occurs at pH 8 or 9, and is greatly depressed by reactions of around pH 5 (Frobisher, 1961). *Mycobacterium tuberculosis* var. grows well but slowly (2-6 weeks) on a solid media such as coagulated egg, serum or blood or on glycerin agar at 37°C. *Vibrio* bacteria (*Streptococcus faecalis*, *Escherichia coli* etc.) also tolerate an alkaline reaction (pH 8-9) (Frobisher, 1961).

Table 1. Effect of different pH on the growth of isolated strains in BCDA a medium

pH of medium	Strains	Incubation period					
		6 h	12 h	24 h	36 h	48 h	72 h
7.1	BCSS(2)	-	+	+++	+++	+++	+++
	BCDump	-	+	+++	+++	+++	+++
	NAIUL	-	+	++	+++	+++	+++
7.6	BCSS(2)	-	+	++++	++++	++++	++++
	BCDump	-	+	++++	++++	++++	++++
	NAIUL	-	-	+	++	+++	+++
9.1	BCSS(2)	-	-	+	+	++	++
	BCDump	-	-	+	+	+	+
	NAIUL	-	-	-	-	+	++
10.6	BCSS(2)	-	-	-	-	-	-
	BCDump	-	-	-	-	-	-
	NAIUL	-	-	-	-	-	-
12.10	BCSS(2)	-	-	-	-	-	-
	BCDump	-	-	-	-	-	-
	NAIUL	-	-	-	-	-	-

aBCDA = Basic Czapek-Dox-Agar; - = No growth; + = Poor growth; ++ = Moderate growth; +++ = Good growth; ++++ = Massive growth.

Table 2. Effect of different pH on the growth of isolated strains in NA[a] medium

pH of medium	Strains	Incubation period					
		6 h	12 h	24 h	36 h	48 h	72 h
4.2	BCSS(2)	-	-	-	-	-	-
	BCDump	-	-	-	-	-	-
	NAIUL	-	-	-	-	-	-
5.7	BCSS(2)	-	-	-	-	+	+
	BCDump	-	-	-	-	+	+
	NAIUL	-	-	-	-	+	+
7.2	BCSS(2)	-	+	++	++	+++	+++
	BCDump	-	+	++	++	+++	+++
	NAIUL	-	+	++	++++	++++	++++
8.7	BCSS(2)	-	-	++	++	++	++
	BCDump	-	-	+	+	++	++
	NAIUL	-	-	+	++	++	+++
10.2	BCSS(2)	-	-	-	+	+	+
	BCDump	-	-	-	+	+	+
	NAIUL	-	-	-	+	+	+

[a]NA = Nutrient Agar; - = No growth; + = Poor growth; ++ = Moderate growth; +++ = Good growth; ++++ = Massive growth.

In this experiment, the bacterial cultures of three strains were incubated at different temperatures like 25, 29, 34, 37 and 40°C. The massive growth of all the strains was found in 37°C. The optimum temperature range for bacteria is from about 25-36°C. A great number of bacteria may grow quite well over the range of 10-40°C (Burges and Raw, 1977). Sultana (1997) observed that 33±4°C temperature was ideal for the growth of bacteria. On the other hand, Sarker (2005) observed that 28.5°C temperature for maximum growth of the isolated strains. Certain bacteria develop most vigorously at temperatures below 20°C. Thermophiles grow well at temperatures of 45-65°C and some thermophiles are in capable of multiplying below 40°C (Alexander, 1961).

Mycobacterium avium grows well at 40°C. *M. chelonei* and *M. marinum* grow well at lower temperatures (18-30°C). *Lactobacillus* sp. grows best on tomato juice agar at 25 to 39°C. *Agrobacterium* sp. grows well on ordinary laboratory media at pH around 6.8 and at temperatures around 25 at 39°C (Frobisher, 1961). *Streptococcus lactis* can be cultivated in sterile milk or on agar containing milk or whey or tomato juice at about 25°C. It grows best in the presence of glucose or lactose. It grows better at about 35°C than at 25°C (Frobisher, 1961).

The strains obtained in this study were incubated at different incubation periods (6, 12, 24, 36, 48 and 72 h). Incubation period 24 h was suitable for well growth of BCSS(2) and BCDump strains, while NAIUL strain was found to be suitable with incubation period of 36 h. Coliform bacteria grow in the incubation period 24±2 h and at 32°C and it shows good growth at 37°C for 48 h of incubation.

Three methods were used (non-microscopic observation or visual observation, microscopic observation, and biochemical tests) to characterize the selected strains. In visual observation, it was found that after 24 hours of incubation, the color of BCSS(2) was light orange, BCDump was white and NAIUL was light brown in their preferred medium (BCDA and NA). After 48-72 hours of incubation, the color of BCSS(2) was orange, BCDump was yellow and NAIUL was brown. The colony type of BCSS(2) and BCDump strains were wet, and NAIUL was creamy.

Staphylococci and *Micrococci* produce golden brown, yellow or white colony on ordinary media. Some enterococci, coryneforms and enterobacteria may produce black colonies on ordinary media. *Staphaylococcus aureus* gives black, shiny, convex colonies on Baird-Parker medium (Collins, 1987).

Gram staining is an old and reliable method for observing the bacteria. Gram negative bacteria were decolorized by alcohol, losing the purple color of crystal violet. Gram positive bacteria did not decolorize and remained purple (Harold, 2002). In the present investigation, all isolated strains were found to be Gram positive. All the isolated strains (BCSS (2), BCDump, and NAIUL) were exhibited positive catalase test because they released oxygen gas from hydrogen peroxide (H_2O_2) by enzymatic degradation. Similar result was observed by UNEP (1996). A number of biochemical tests have been performed to study the characteristics of bacterial strains (Table 3). Fermentation test is used to differentiate the microorganisms that ferment carbohydrate (such as lactose, mannitol). The isolated strains exhibited positive lactose and mannitol fermentation test because open and sealed tube produced yellow color.

In urease test, due to the formation of red pink color through the media, BCDump strain showed positive urease test and this strain was able to decompose urea to ammonia. BCSS(2) and NAIUL showed negative test. Because they did not produce red-pink color in this the media.

Table 3. Biochemical tests of some waste decomposing bacteria

sniartS	Catalase	Lactose Fermentation Test	Mannitol Fermentation Test	Voges-Proskauer Test	Urease Test	Indole Test	Citrate Utiliza tion Test	Identified Bacteria
NAIUL	Positive	Positive	Positive	Negative	Negative / Positive	Negative/ positive	Positive	*Xanthomonas* spp.
BCDump	Positive	Positive	Positive	Positive	Positive	Negative	Negative	*Bacillus* spp.
BCSS (2)	Positive	Positive	Positive	Positive	Negative	Negative	Negative	*Pseudomonas* spp.

Indole test demonstrates the ability of certain bacteria to split the amino acid tryptophan to indole which accumulates in the medium. The isolated strain NAIUL showed positive test and produced red color in media. BCDump and BCSS(2) strains showed negative test because they did not produce red color in media. Sultana (1997) also observed the same result.

BCDump and NAIUL strains showed positive test in hydrogen sulphite production test because they produced black color in media. However, BCSS(2) displayed negative test as this strain did not produce black color in media. The same result was observed by Sultana (1997).

Citrate utilization test was based on the ability of an organism to use citrate as its only carbon source and ammonia as its only source of nitrogen. The isolated strain NAIUL showed positive citrate utilization test because media turned green to blue. BCSS(2) and BCDump did not turn the media green to blue and therefore, they showed negative test (Table 3).

Studies were conducted by various workers and the results showed accumulation and separation of lower volatile fatty acids and VOC_s (Volatile Organic Compounds) during domestic waste decomposition (Kaplovsky, 1952). As the waste breakdown by microorganisms (bacteria), the weight and volume of the matter decrease. In the present study of decomposition, we observed that the weight and volume of treated garbage were decreased. Because bacteria used the garbage and broken down it and converted to simple molecules.

For solid waste degradation (without additive), the highest weight loss in suspension treatment was recorded 50.70 % after 30 days by treating with BCDump strain. In similar study, Hoitink, (1980) found the highest weight loss (25.76 %) using *Trichoderma* strain after 30 days. The highest volume loss in suspension treatment was found 38.67 % for BCDump strain after 30 days. In case of pellet treatment, the highest volume loss was 37.53 % in BCDump strain after 30 days.

Additives such sucrose can be used with garbage for decomposition which can increase the growth of bacteria (Collins, 1987). It was observed that by treating with sucrose solution, the growth of bacteria was high and decomposition rate was also high. It seems that addition of sucrose solution to the garbage enhanced decomposition rate through increasing fermentation. In 5 %, 10 % and 15 % sucrose solution treatment (along with bacterial strain), the highest weight losses percentages were found 52.73 %, 55.32 % and 65.12 %, respectively in BCDump strain after 30 days of inoculation. Among those treatments, the highest weight loss was 65.12 % in BCDump strain along with 15 % sucrose solution. Similar trend was found for volume loss. The highest volume loss was found 51.34 % after 30 days of inoculation in BCDump strain along with 15 % sucrose solution. So it can be inferred that in sucrose treatment, the most effective strain was BCDump along with 15 % sucrose solution (in term of weight loss and volume loss).

As sucrose treatment was effective for garbage decomposition, in later in order to reduce the cost, molasses were used as the alternative and cheap source of carbohydrate for decomposition of solid waste. In 5 %, 10 % and 15% molasses solution treatment, the highest weight loss percentages 56 %, 62.70 % and 80.24 %, respectively were found in BCDump strain after 30 days inoculation. Among those treatments, the highest weight loss 80.24 % was found in 15 % molasses solution in BCDump strain. Similar trend was found for volume losses. The highest volume loss was found 64.27 % after 30 days of inoculation in BCDump strain along with 15% molasses solution (in term of weight loss and volume loss). So it can be inferred that in molasses treatments, the most effect strain was BCDump along with 15 % molasses solution (in terms of weight loss and volume loss).

Again comparing the effect between different concentrations of sucrose and molasses treatment for degradation of solid waste with bacterial suspension, the highest weight loss was 80.24 % in 15 % molasses solution in BCDump strain, whereas the highest weight loss 65.12 % was found in 15 % sucrose solution in BCDump. Similar trend was found for volume losses. Therefore, it can be concluded that the garbage decomposition by 15%

molasses was the most effective followed by 15 % sucrose solution in term of volume losses, whereas bacterial suspension (without any additive) was able to lower volume change than additives treatments. In 15 % sucrose treatment, the highest volume loss (%) observed 51.34 % after 30 days of inoculation in BCDump strain and in 15% molasses solution, the highest volume loss (%) was 64.27% after 30 days of inoculation in BCDump strain. In similar study, Bari (2006) found 66 % weight loss for *Trichoderma* strain in 4 % concentration of glucose level. Harper *et. al.,* (2001) reported that in composting pile, the final weight of 26.1 kg (58 lb) of afterbirth and dead piglets after composting for two weeks was only 3.1 kg (6.9 lb), and the remaining tissue was easily crumbled in the sawdust medium. In this experiment, the average daily weight loss was more than 6 % of the original animal mass.

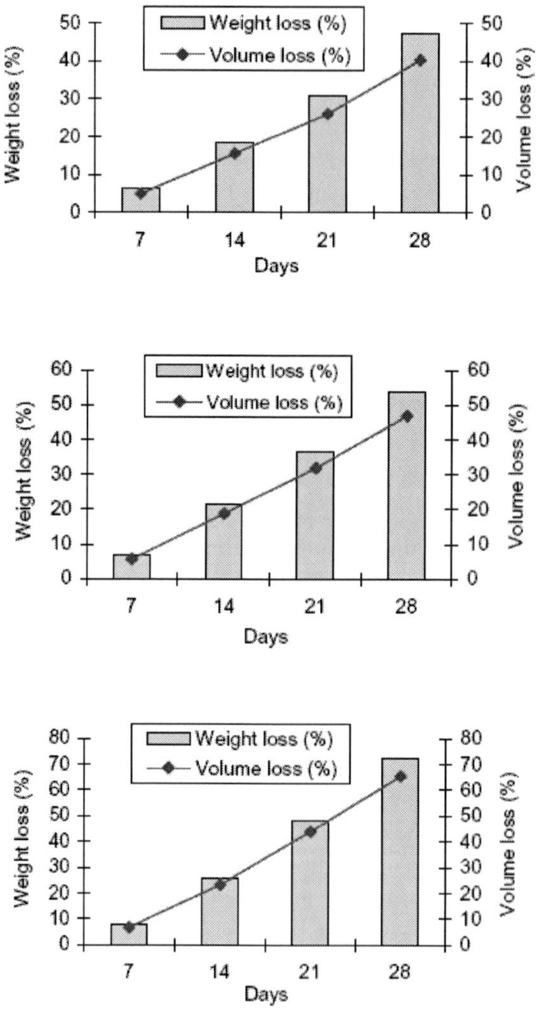

(A)

(B)

(C)

Figure 1. Weight loss (%) and volume loss (%) of decomposed garbage at 7 days interval using BCDump suspension along with different concentrations of molasses solution.
(A), BCDump + 5% molasses solution; (B), BCDump + 10% molasses solution;
(C), BCDump + 15% molasses solution.

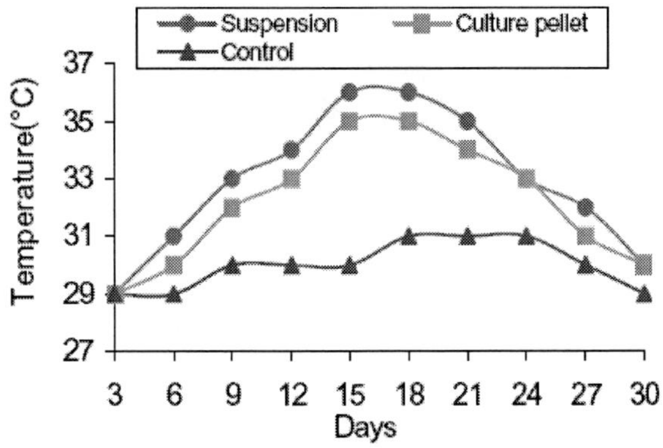

(A)

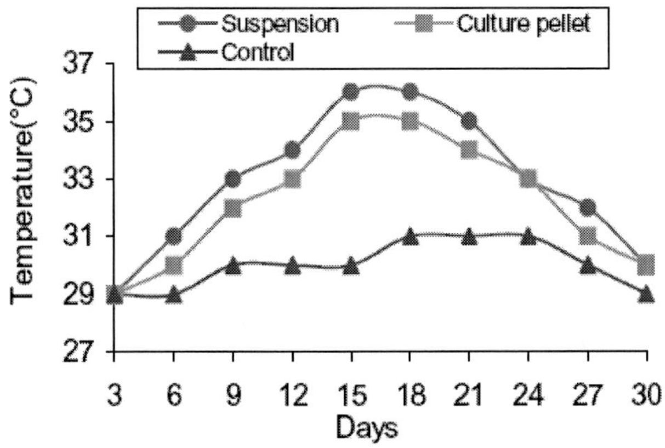

(B)

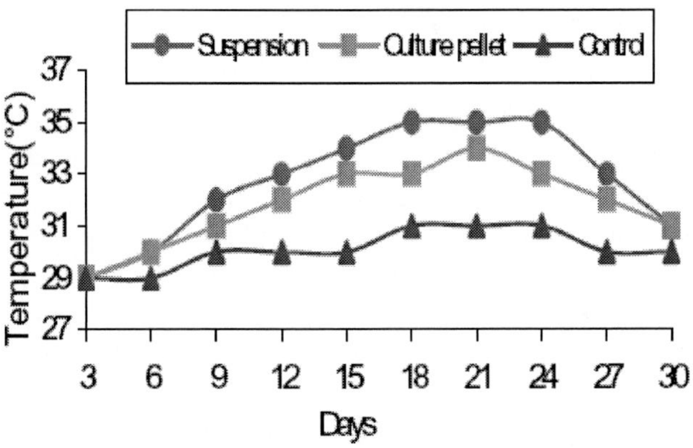

(C)

Figure 2. Changes of temperature (°C) during garbage decomposition at 3 days interval using suspension and culture pellet of the three strains. (A), BCDump strain; (B), BCSS(2) strain; (C), NAIUL strain.

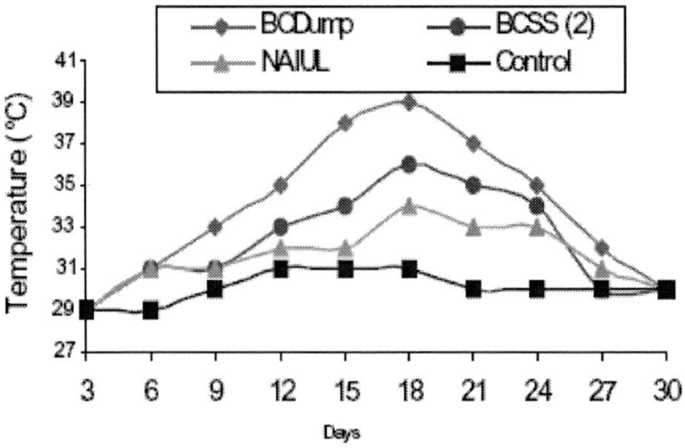

(A)

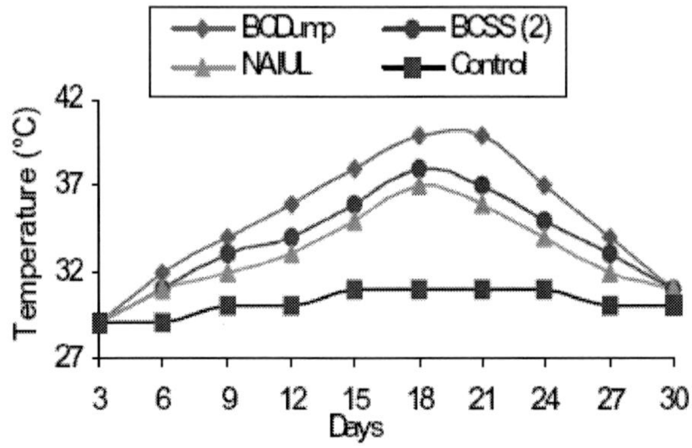

(B)

Figure 3. Changes of temperature (°C) during garbage decomposition at 3 days interval using the three strains separately along with the most effective concentration of additives. (A), three strains + 15% sucrose solution; (B), three strains + 15% molasses solution.

In the present investigation, it was observed that in term of weight loss (%) and volume loss (%), molasses treatment was better than sucrose treatment to decompose organic solid waste with bacterial suspension. It can be explained in this way that probably in molasses, there were some extra components (which were absent in sucrose) that could give extra nutrient to rapid growth of bacteria and to increase the number of bacterium cell. So the degradation process with molasses solution was higher with higher number of bacteria than sucrose solution treatment. So, the growth of bacterial strain was the main factor to decompose garbage.

It was observed that the relationship between weight loss and volume loss was linear and no interaction was found between the two parameters. Both volume loss (%) and weight loss (%) were increased gradually with the progression of decomposition process and the result was same for each treatment of each strain. The weight loss (%) was higher than volume loss (%) in garbage decomposition in each treatment of each strain as shown in Figure 1.

When microbes degrade waste, heat is produced and temperature is increased with decomposition process (Moqsud, 2003). In this study, we observed that in all cases temperature gradually increased after 4-6 days, and the temperature reached at the peak after 15-24 days and declined gradually after 28-30 days, and reduced to a level as it was in its initial stage. For decomposition (in composting method) of solid waste in *in vitro* condition by bacterial suspension and culture pellet, the highest temperature was 37°C after 15-18 days in suspension treatment of BCDump and BCSS(2) (Figure 2 and 3).

Using sucrose solution, the highest temperature was 39°C after 18 days in BCDump strain along with 15 % sucrose solution. For molasses solution, the highest temperature was 40°C after 18 days in 15 % molasses solution in BCDump strain. So it can be inferred that high decomposition rate is possibly associated with temperature. In this investigation, the weight and volume loss were the highest in which treatment the temperature was found to be the highest. In all cases, the temperature was high in BCDump strain and the weight loss and volume loss were also the highest in this strain.

In this investigation, the initial pH of fresh garbage was 4.11 (acidic). In all cases, the pH of decomposed garbage (in suspension and pellet treatments, sucrose and molasses treatment) became alkaline from its initial acidic state, ranged from 7.31-11.06. The highest pH 11.06 was noted using 15% molasses solution decomposed by BCDump strain.

Table 4. Changes of Temperature (°C) during garbage decomposition by bacterial suspension and different concentration of molasses solution at 3 days interval

Days	Temperature (°C) of decomposition garbage by treating BCDump strain and molasses solution				Temperature (°C) of decomposition garbage by treating BCSS(2) and molasses solution				Temperature (°C) of decomposition garbage by treating NAIUL strain and molasses solution			
	5% MS[a]	10% MS[a]	15% MS[a]	CT[b]	5% MS[a]	10% MS[a]	15% MS[a]	CT[b]	5% MS[a]	10% MS[a]	15% MS[a]	CT[b]
3	29	29	29	29	29	29	29	29	29	29	29	29
6	31	31	32	29	31	31	31	29	31	31	31	29
9	33	33	34	30	33	33	33	30	32	32	33	30
12	35	35	36	30	34	35	35	30	33	33	34	30
15	37	37	39	30	36	37	37	31	35	35	36	31
18	37	38	40	31	36	37	38	31	35	36	37	31
21	35	38	40	31	34	35	37	31	33	34	36	31
24	33	36	37	31	32	34	35	31	31	32	34	31
27	32	34	34	30	31	32	33	30	30	31	31	30

[a] MS, Molasses solution, [b] CT, Control.

When bacteria are cultivated in a medium, pH 7 is very likely that this pH will change as a result of substance produced by the organism, which may be either acidic or basic (Pleczar *et. al.*, 2004). It was reported that the pH varies with time during the composting process. The optimum pH range for most bacteria is between 6.0 and 7.5. During the initial period (first 2 to 7 days) pH drops to 6 or less and then begins to rise to about 7.5-8.5 for the remainder of the composting. In the initial stage of composting, organic acids are produced. For this acidic condition the pH value reduces in the initial days. But after some days, the temperature rises and the more decompositions of organic waste cause the rise in pH value (Jackson, 1974). For decomposition of organic kitchen wastes by bacteria culture pellet, the odor was bad smell after 10, 20 and 30 days that indicate possible slow degradation of organic matters. In decomposition of solid waste by bacterial suspension, it was found that after 10 days the odor was bad smell and after 30 days there was no smell that indicates the possible complete degradation of organic kitchen wastes. In similar study, Rahman (2004) reported that, in decomposition of solid waste by composting the odor was reduced. The study suggests that the useful bacteria might be isolated from the surrounding environment for friendly bioconversion of solid organic waste. To protect human health and natural environment from waste problems, the developed method of decomposition of waste as shown in this study is found to be very effective and useful. The study was conducted to observe the effect of fresh garbage, decomposed garbage and bacterial suspension on biomass production (fresh weight basis) of potato (*Solanum tuberosum*). For conducting the experiment, decomposed garbage (using bacterial suspension and 15 % molasses), fresh garbage, and bacterial suspensions were used separately in the soil pot of potato. Control treatment (only garden soil) was also performed for each treatment. It was observed that decomposed garbage treatment (decomposed by BCDump suspension and 15% molasses) showed highest biomass production (65.61g), whereas fresh garbage and bacterial suspension exhibited 50.98 and 39.31 g, respectively. Decomposed garbage treatment (decomposed by BCSS(2) suspension and 15 % molasses) showed the highest biomass production (58.41g), whereas fresh garbage and bacterial suspension exhibited 48.45 and 37.07 g, respectively. It was observed that decomposed garbage treatment (decomposed by NAIUL suspension and 15% molasses) showed the highest biomass production (52.18 g), while fresh garbage and bacterial suspension exhibited 47.69 and 34.19 g, respectively. It was found that in all cases, addition of decomposed garbage in soil enhanced biomass production when compared with fresh garbage or bacterial suspension treatment. The highest biomass in term of fresh weight achieved using decomposed garbage by BCDump suspension along with 15 % molasses solution followed by BCSS(2) and NAIUL. In all cases, fresh garbage and bacterial suspension performed better than control (garden soil only).

CONCLUSION

From this present investigation, it can be concluded that the beneficial bacteria such as *Xanthomonas* spp., *Bacillus* spp., and *Pseudomonas* spp. isolated from the surrounding environment and identified by subsequent biochemical tests could be used effectively for degrading the solid waste and the degraded substences could be used as organic manure or compost. Therefore, boidegradation of solid waste by microorganisms represents an alternative method for disposal of biodegradable portion of the waste material for organic manure and to protect environment contamination and health hazards from waste materials.

REFERENCES

Arlosoroffs,(1982). WB/UNDP Integrated resource recovery project: ‘Recycling of wastes in developing countries’, In: *Appropriate waste management for developing countries’*, edited by Kriton Curi. New York: Plemum Press, 1985.

Alexander, M. (1961). '*Introduction to soil microbiology*', John Wiley and Sons, Inc, New York

Bangladesh Bureau of Statistics (BBS). (1997). 'Bangladesh population census 1991', *Urban Area Report Dhaka: Statistics Division*, Ministry of Planning and Government of Bangladesh.

Blight, G. E. & Mbande, C. M. (1996). 'Some problems of waste management in developing countries', *Journal of Solid Waste Technology and Management,* Vol. 23(1), pp.19-27.

Burges, A. & Raw, F. (1977). '*Soil biology'*, Academic Press. Inc., 111 Fifth Avenue, New York.

Berkeley, R. C. W. & Campbelt, R. (1972). Microbial nutrition and the influence of environmental factors on microbial growth and other activities. In: *Microorganism: function, form and environment.* Hawker, L.E and Linton, A.H. (ed), Edward Arnold (Publisher limited). London, England

Collins, C. H. & Lyne, P. M. (1987). '*Microbiological methods'*, 5th ed., Butterworth & Co. Ltd., London: 88 Kings way, England.

Cointreau, S. (1982). 'Environmental management of urban solid wastes in developing countries', *A project guide*, Washington, DC: Urban Development, World Bank.

Frobisher, M. (1961). '*Fundamentals of microbiology',* Charles E. Tuttle, 15 Edogawa-cho, Bunkyo-ku, Tokyo, Japan.

Harold, J. B. (2002). *'Microbiological applications',* Laboratory manuals in general microbiology, 8th edition. McGraw-Hill Higher Education.

Harper, A. F., Estienne, M.J. & Collins, E. R. (2001). Composting as an environmentally safe means of dead pig disposal on Virginia swine farms. Suffolk, Virginia: Virginia Tech Tidewater Agricultural Research & Extension Center. Retrieved June 16, 2003, Url:http://vmirl.vmi.edu/ev/Paper%20Sessions/Early%20Bird/Absracts/Harper_Abstract.htm.

Hoitink, H. A. J. (1980). 'Composted Bark, A lightweight growth medium with fungicidal properties', *Plant Disease,* Vol. 64, pp.142-147.

Kaplovsky, A. J. (1952). Volatile acid production during digestion of several industrial wastes, *Sewage and Ind. wastes,* 24, 194-201.

Moqsud, M. A. (2003). 'A study on composting of solid waste', *M.Sc. Eng. Thesis*, Department of Civil Engineering, BUET, Bangladesh.

National Institute of Local Government (NILG) & Local Government Engineering Department. (2002). *Pouroshava Statistical Yearbook of Bangladesh*, 1998-99, NILG: Dhaka.

Pleczar, M. J., Chan, E. C. S. & Krieg, N. R. (2004). '*Microbiology',* Tata MCGraw-Hill Puplishing Co. LTD, New Delhi, India.

Paimela-Wheler, A. (2004). World Bank financed waste management projects: Overview of the waste management markets and trends in developing countries and prospects for finish companies. Project Advisory Services, Washington

Parr, J. F. & Hornick, S. B. (1992). 'Utilization of municipal wastes', In: Soil microbial ecology: *Applications in agricultural and environmental management,* F.B. Metting (ed.) Marcel Dekker, Inc., New York, U.S.A. pp. 545-559.

Rahman, M. H. (2004). 'Composting of solid waste in Bangladesh', Proc. of the 19th International Conf. On *Solid Waste Tech. and Manage.*, Philadelphia, USA.

Sustainable Development Networking Programme (SDNP), (2005). '*World Environment Day 2005*', URL: http:// www. sdnpbd. org/sdi/ international_ days/wed/2005/index.htm.

Sultana, S. (1997). *Isolation of cellulolytic microorganisms and their activities*. M. Phil. Thesis. Institute of Biological Science. University of Rajshahi.

Tchan, Y. T. (2002). *Proc. Linn. Soc. N.S.W*, 77, 89-91.

United Nations Environment Programme (UNEP). (1996). '*International book on environmentally sound technologies for municipal solid waste management*', UNEP Technical Publication 6.

United States Environmental Protection Agency (USEPA). (1989). *EPA'S policy promoting the beneficial use sewage sludge; and the new proposed technical sludge regulations.* Office of Municipal Pollution Control, Washington, D.C., USA

Zurburgg, C. (2002). '*Urban solid waste management in low-income countries of Asia how to cope with the garbage crisis*', Scientific Committee on

Problems of the Environment (SCOPE) urban solid waste management review session, Durban, South Africa.

In: Characterization and Development of Novel Materials ISBN: 978-1-61324-149-3
Editor: Rafiq Islam

Chapter 12

KARANJ: A POSSIBLE SOLUTION FOR THE RURAL ENERGY AND OTHER USES

M. P. Islam*, M. A. Haque and M. D. Hussain
Department of Farm Power and Machinery,
Bangladesh Agricultural University, Mymensingh, Bangladesh

ABSTRACT

This paper has discussed the oil potential of Karanj (Pongamia pinnata) as source of bio-diesel, for cultivation in Bangladesh and other potential uses. This paper has also outlined the oil extraction method from the seed which is s chemical solvent method known as cold percolation method by using the CCl4 and this method can extract more than 95% oil. Physical properties of the Karanj seed are also discussed and the bulk density of the seed is 0.56g/cc. Karanj seed contains oil about 32%. The specific gravity of Karanj oil varies from 0.925 to 0.940 at 200 C which is slightly higher than the sp. gravity of fossil diesel. Therefore, it should be adjusted for its large scale application in engine running. The high viscosity of Karanj oil interferes with injection process and leads to poor fuel atomization. The injection system may be slightly modified for the application of Karanj oil as fuel for diesel engine or the high viscosity has to be overcome by using methyl ester with Karanj oil. The transformation of Karanj oil to its methyl ester reduces molecular weight to one third, reduces viscosity to one eighth and increases the diesel index (Anonymous, 2008a). The flashpoint of the oil is 134 °C which explain that the Karanj oil could stand for a long time. So, Karanj cultivation in Bangladesh could be a practical solution for the rural people in order to minimize the energy shortage.

Keywords: Karanj oil, oil extraction, rural energy, bio-diesel, Physical, chemical and thermal properties

* Corresponding author E-mail: engr_parvezfpm@yahoo.com.

1. INTRODUCTION

Karanj is a medium sized tree that generally attains a height of about 8 m within 5 years and a trunk diameter of more than 50 cm (Figure 1). The trunk is generally short with thick branches spreading into a dense hemispherical crown of dark green leaves. The bark is thin gray to grayish- brown, and yellow on the inside. The taproot is thick and long; lateral roots are numerous and well developed. It is widely grown in the hilly areas of Bangladesh. The most important character of this plant is the nitrogen fixing capability. It is often planted as an ornamental and shade tree but now-a-days it is considered as an alternative source for bio-diesel. No pests are of major concern, but caterpillars occasionally cause some defoliation. Native to humid and subtropical environments, karanj thrives in areas having an annual rainfall ranging from 500 to 2500 mm (Satish, 2007) in its natural habitat, the maximum temperature ranges from 27 to 38° C and the minimum 1 to16° C. This species grows to elevations of 1200 m, but in the Himalayan foot hills is not found above 600 m (Satish, 2007). It can grow on most soil types ranging from stony to sandy to clay. The drought tolerance capability of this tree is high but does not grow well on dry sands. Salt tolerant capability is moderate. It is common along waterways or sea shores, with its roots in fresh or salt water. Highest growth rates are observed on well drained soils with assured moisture. Natural reproduction is profuse by seed and common by root suckers. It will show nutritional deficiencies if grown on soil with a pH above 7.5 (Edward *et al.* , 1994).

Renewable energy such as Karanj oil which has similar characteristics like diesel oil, could be an alternative solution to fulfill the requirements of the cheaper source of energy in the rural areas of Bangladesh.

Figure 1. Karanj tree at BAU campus.

2. METHODOLOGY

The methodology applied to this study has been discussed here.

2.1. Agro-practices - Pit Preparation

The size of the pits where the cuttings or seedlings were planted was 50 cm * 50 cm. The soil was taken out and kept open for one week to sundry. While refilling the pits stones and boulders were removed and filled with 1/3 normal soil, 1/3 sand and 1/3 compost.

2.2. Propagation Method

Direct seeding can be successful under well prepared soil and optimal moisture content and using better quality of seeds from healthy mother plant.

2.3. Harvesting

The fruits were collected from the branch by hand picking. The harvested fruits were kept in dry place.

2.4. Pruning

Pruning is an important maintenance practice for Karanj. Because pruning is necessary for the proper growth of Karanj tree and thus enhance the yields too. The best time for general pruning is in late winter or early spring, immediately before growth resumes.

2.5. Management

The plant does well when rich organic nutrition was provided. Plants show significant increase of production if fertilizing regularly. Hoeing and weeding at least twice a year is necessary, especially during the establishment period. Weeding was accomplished as and whenever necessary to keep the plant free from weeds for better soil aeration and to break the crust. It also helped in soil moisture conservation.

Almost every type of oil seed requires drying to minimize the moisture level available in the seed for storage and processing. After collection of the fruits they were dried but most remember that direct sun has a negative effect on seed viability and seeds should be dried in the shade. The seeds were dried after they were separated from the fruits and cleaned. Seeds must be stored in air tight containers and should not be stored in damp and dark place.

2.6. Physical Properties

Size, shape, weight, density, volume are some of the physical characteristics which are important in many problems associated with design of a specific machine or analysis of the behavior of the products. Size (length & width), Weight, bulk density and moisture content of the seeds (Figure 2, Left) were measured at the Processing Lab of the department of Farm Power & Machinery, Bangladesh Agricultural University. The weight of the selected seeds (each of 20 numbers) were measured by a digital balance (TC-203, Denver Instrument Company, USA, with a readability of 0.001 g), the average value of seeds were tabulated with standard deviation. The size (length and width) of the seeds was measured by a slide caliper and the bulk density was measured by volumetric method using a measuring cylinder of 1000 ml (Figure 2) and a digital balance. Bulk density was expressed in *g/cc*. Moisture content of the seeds was determined by oven dry method at 105^0C for 24 hours. The moisture content was calculated in dry basis.

2.7. Mechanical Properties

The seed hardness was measured at BRRI (Bangladesh Rice Research Institute) by using a hardness tester (Kiya Seisakusho Ltd. Tokyo, Japan) (Figure 2, Right). When the seed started to crack then the reading of the load indicator was observed. The crushing strength was calculated with respect to the projected area of the indenter used. The diameter of the indenter was 3 mm.

$$Crushing\ strength\ ,(kg/cm^2) = \frac{Weight\ required\ cracking\ the\ seed\ ,kg}{Projected\ area\ of\ seed\ underload\ ,cm^2}$$

Figure 2. Measurement of the seed bulk density (left) and hardness (Right).

Figure 3. Oil extraction by the cold percolation method.

2.8. Oil Extraction

Oil can be extracted from the seeds by applying heat, solvents or by pressure. Extraction by heat is not used commercially for vegetable oils. The oil from seeds can be extracted by three different methods. These are mechanical extraction using a screw press, solvent extraction and cold percolation method. In this experiment oil percentage was estimated by cold percolation method (Gravimetric method) in laboratory (Figure 3), (Kartha et al., 1957) is described below. The word cold in this context means no heat was applied and extraction occurs at room temperature.

During the use of CCl_4 hazardous solvents, following safety precautions were taken in consideration:

Wipe up all spills immediately.

Place rags or other items used to clean up spills in a separate, covered container.

Use protective clothing, goggles, gloves, or other appropriate safeguards to prevent solvents from getting on skin or in eyes.

Have accessible fire-fighting equipment nearby.

Have adequate ventilation.

2.8.1. Preparation of seeds, equipment and chemicals used for oil extraction

Matured seeds were collected from ripe fruits and were sun dried. They were decorticated manually. To prepare the seeds for oil extraction, they should be solar heated for several hours or roasted for 10 minutes. The seeds should not be overheated. The process breaks down the cells containing the oil. The heat also liquefies the oil, improves the extraction process.

Agate mortar and pestle, percolator and sintered glass funnel, shaking machine, 20 ml air tight plastic bottle, 100 ml beaker, sand bath, Mattler electronic balance, Sodium sulphate (Na_2SO_4), Carbon tetrachloride (CCl_4) and finely crushed glass powder.

2.8.2. Procedure

The sample weight of the seed powder was 0.3g, which was taken into a mortar with 2g of glass powder and Na_2SO_4. The mixture was grinded to a fine stage. Then 10 ml of CCl_4 was put into the mortar to make a solution. The solution was then taken into a 20 ml vial. The volume of the solution was made 20 ml by adding CCl_4. The vial was shaken over night by using a shaker. The solution was then filtered with the help of sintered glass funnel and percolator. The filtrate was collected in a pre-weighed beaker with two glass balls. The beaker containing filtrate (oil + CCl_4) with two glass balls wwas placed on a sand bath for evaporation of CCl_4. Evaporation was done at 60-70^0 C in a sand drier. After the evaporation the beaker containing oil was kept in desiccators for cooling. Finally the beaker with oil and glass balls was weighed accurately.

2.8.3. Calculation of oil content

$$\% \text{ of oil} = \frac{(w_2 - w_1) \times 100}{w}$$

Where,
w = weight of sample, g
w_1= weight of the beaker with glass ball, g
w_2 = weight of the beaker with glass ball and oil, g

3. AGRONOMICAL PRACTICES

3.1. Botany

Scientific name *Pongamia pinnata (*family *Leguminosae*, subfamily *Papilionoideae)*, whose local name is Karanj.

Figure 4. Leaves (a), Flowers (b), Fruits (c) and Seeds (d) of karanj at BAU campus (Islam, 2008).

3.2. Leaves

Leaf elliptic (oval) shaped, alternate leaf arrangements (Figure 4a), evergreen and leaflet blade length is 5 to 10 cm and odd pinnately compound type (Edward et al., 1994).

3.3. Flower

Flowers purplish or whitish (Figure 4b), contains 2 - 4 petals. Flowering time throughout the year but May to June and August to September is the main season of flowering.

3.4. Fruit and Seed

Fruits length are 2.5 to 7.5 cm, dry or hard covering and brown colour (Edward et al., 1994). Pods are elliptical, 3-6 cm long and 2-3 cm wide, thick walled, and usually contain a single seed (Figure 4c, d). Seeds are 1-2 cm long, oblong, and light brown in color. The kernels are tough white in colour, covered with thin reddish brittle skin. In dry pods, the shell and kernel are 4.5% and 95.5 %, respectively. The composition of a sample of air dried kernels in percent is shown below (Kureel et al., 2007):

Components of Kernels:

Moisture	19.0
Oil	at least 27.5
Protein	17.4
Starch	6.6
Crude Fibre	7.3
Ash	2.3

3.5. Trunk/Bark/Branches

Droop as the tree grows, and will require pruning for vehicular or pedestrian clearance beneath the canopy; routinely grown with, or trainable to be grown with, multiple trunks; not particularly showy; tree wants to grow with several trunks but can be trained to grow with a single trunk; no thorns (Islam, 2008).

3.6. Distribution

The natural distribution of Karanj is along coasts and river banks in India, Burma and the tidal forests, along river and canal banks, wetlands of haor basin in the northeast region of Bangladesh. Native to the Asian subcontinent, this species has been introduced to humid tropical lowlands in the Philippines, Malaysia, Australia, the Seychelles, the United States and Indonesia.

3.7. Seed Oil Extraction

A thick yellow-orange to brown oil is extracted from seeds. Yields of 20 to 25% of volume are possible using a mechanical expeller (Satish, 2007). But by using solvent such as CCl_4 the higher percentage of oil is possible to extract which is approximately 32%.

Table 1. Physical, chemical and thermal properties of the Karanj oil (Anonymous, 2007c and Anonymous, 2007b)

Attributes	Property/Volume
Appearance	Pale yellow liquid
Colour	Dark brown
Odour	Characteristic odd odor
Solubility	Soluble in alcohol and oils. Insoluble in water
Specific Gravity	0.925 – 0.940 @ 20°C
Viscosity at 40 °C,mm^2/s	27.84
Iodine value	80 - 96
Saponification Value	185 – 195
Unsaponifiables	2.6 - 3.0 %
Acid value, mg (KOH/g)	5.06
Palmitic Acid	3.7 – 7.9
Stearic Acid	2.4 – 8.9
Oleic Acid	44.5 – 71.3
Linoleic Acid	10.8 – 18.3
Extraction Method	Cold-pressed of seeds or chemical method.
Contents	Palmitic acid, stearic acid, oleic acid, linoleic acid, arachidic acid, gadoleic acid, behenic acid, lignoceric acid, eicosenic acid and alkaloids – karangin, kanugin, kanjone, neoglabrin, pinnatin, glabrin, pongamol, pongapin, quercitin, demethoxy-kanugin, glabrosaponin, gamatay, saponin, kaempferol, β-sitosterol and tannin.
Cloud Point	13.2
Flash Point	134 °C
Calorific value,kJ/Kg	37,304
Cetane Value	55
Ash content, %	0.014
Carbon residue, %	1.51
Fire point, °C	209
Fire/Explosion Hazard:	
Flash Point: 40 °C Flammable liquid. Will ignite if moderately heated.	

4. USES

4.1. Wood

With a calorific value of 4600 - 4800 kcal per kg, karanj is commonly used as fuel wood. Its wood is beautifully grained and medium to coarse textured (Antrim, 2008). However, it is not durable, is susceptible to insect attack, and tends to split when sown. Thus the wood is not considered a quality timber. The wood is used for cabinet making, cart wheels, posts, agricultural implements, tool handles and combs (Satish, 2007).

4.2. Oil

The Karanj oil has a bitter taste and a disagreeable aroma, thus it is not considered edible. The oil is used as a fuel for cooking and lamps in the rural area. The oil is also used as a lubricant, water-paint binder, pesticide, and in soap making and tanning industries for dressing of leathers. The oil is known to have value in folk medicine for the treatment of rheumatism, as well as human and animal skin diseases. It is effective in enhancing the pigmentation of skin affected by leucoderma or scabies (Satish, 2007). The oil of Karanj is also used as a substitute for diesel and properties are shown in Table 1.

The colour of the Karanj oil is dark brown and it is soluble in alcohol and oils. But insoluble in water. Specific gravity of the oil is 0.925 – 0.940 at 20^0 C. Acid value of the Karanj oil is 5.06 mg (KOH/g) and calorific value is 37,304 kJ/Kg with carbon residue 1.51% which demonstrate that this oil could be a good substitute of fossil diesel oil.

Oleic acid is the principal unsaturated fatty acid averaging 45 to 70% of all the commercially available oils from tree borne oil seeds like neem and sal, karanj containing maximum oleic acid, is a valuable raw material for oleic acid derived oleo chemicals apart from other traditional uses. The oil can be used after neutralisation as a lubricant for heavy lathes, chains, bearings of small gas engines and heavy engines. The oil and its active component karanj possess insecticidal and antibacterial properties, 2% karanj oil resin soap spray is reported to be effective against the nymph and adult stages of the green bug of coffee (Kureel *et al.*, 2007).

4.3. Fodder and feed

Opinions vary on the usefulness of this species as a fodder. The leaves are eaten by cattle and readily consumed by goats. However, in many areas it is not commonly eaten by farm animals. Cake is rich in protein (30-35%) but it is unpalatable and toxic to cattle even at 4% level due to presence of several toxic particles (karanjin, pongamol etc.). These toxins are oil soluble and most of the toxins are removed during solvent extraction of oil from cake. Therefore, de-oiled cakes could be used in compound cattle feed (Kureel et al., 2007).

4.4. Other Uses

Dried leaves are used as an insect repellent in stored grains. The oil cakes when applied to the soil have pesticidal effect, particularly against nematodes and also improve soil fertility. Karanj is often planted in homesteads as a shade or ornamental tree and in avenue plantings along roadsides and canals. It is a preferred species for controlling soil erosion and binding sand dunes because of its dense network of lateral roots. Its root, bark, leaf, sap, and flower also have medicinal properties.

The cake contains nitrogen (N) 5.1%, phosphorous (P_2O_5) 1.1% and potash (K_20) 1.3% and has been reported as a useful organic manure for sugarcane, coffee, oranges and paddy. It repels red ants and also reported effective for nematode controls. It also possesses nitrification inhibition property and cake is used in blending with nitrogenous fertilizers (Kureel et al., 2007).

5. Agro-Practices

5.1. Propagation

Karanj is easily established by direct seeds or by planting nursery-raised seedlings or stem cuttings of 1-2 cm root-collar diameter. In peninsular India, the seeding season is April to June, and the seed yield per tree ranges from about 10 kg to more than 50 kg (Kureel et al., 2007). But in Bangladesh, the seeding season is June to July and September to October. Approximately 1500-1700 seeds weigh one kilogram.

5.2. Seed Collection and Storage

The pods are collected from May to July. The pods are dried in sun and seeds are extracted by thrashing the fruits. The natural longevity of seeds is 6 months. They remain viable for a year when they are stored in air-tight containers at 5^0C (Kureel et al., 2007). The seeds can be stored with the fruit shell unopened in sealed polythene bags.

5.3. Germination

Seeds are sown in polybags with the micropyle facing downwards. Seeds start germinating from 10th day of sowing. Generally, seeds do not require any pretreatment before sowing. But soaking the seeds in hot water (50^0 C) for 15 minutes improves germination percent and also vigour (Kureel et al., 2007). The germination in this species is hypogeal i.e. in normal seedling, the cotyledons remain beneath the germination medium while the plumule pushes upwards and emerges above. Seedlings attain a height of 25-30 cm in their first growing season (Satish, 2007).

5.4. Raising of Seedlings

The seedlings are raised in seed beds or polypots. Seed is sown in previously prepared raised seedbeds. The seed is dibbled in the nursery beds since the beginning of hot weather at a spacing of 7.5 cm x 1.5 cm. Mulching of beds is helpful. In some places seeds are sown in polypots during July to August. The seeds start germinating 10 days after sowing and germination will be complete in about a month (Kureel *et al.*, 2007). When seedlings are 50 cm in height, transplanting to the field should occur at the beginning of the next rainy season. Seedlings have large root systems. So, soil should be retained around the roots during transplantation. The spacing adopted in avenue plantings is about 8 m between plants and in block planting, the spacing can range from 2 m x 2 m to 5 m x 5 m or in the formation of 5 m x 4 m, it is possible to plant 200 tress/acre (Satish, 2007).

5.5. Pruning

For developing the strong structure pruning requires once in every rainy season from the 2nd year. Because more the branches, more the production of fruits and seeds.

5.6. Harvesting

The possible yield is around 20 to 45 kgs per tree. One person can collect 180 kgs of seeds in 8 hours per day (Satish, 2007). Seed collection season in Bangladesh is from December to April.

5.7. Management

It's a relatively low maintenance tree once established, which has high resistant capacity against winds, drought and salinity. The optimum amount of organic manure is about 2 kg along with chemical fertilizers containing N (30 g), P (60 g) and K (20 g) mixed with soil in the pit (50 cm x 50 cm) and applied at the time of planting and then once in every 6 months (Kureel *et al.*, 2007). It is very important to cut the tree in time and keep it in proper shape. The plants should be cut in proper way, so that it will grow like umbrella. Care should be taken from the beginning.

5.8. Economics

Large scale jobs can be generated from the creation of nurseries, plantation and maintenance of the trees, harvesting and processing of seeds. It is the most fast growing tree available in Bangladesh on not only for possible source of biofuel but also for biomass. The seeds yield the oil which is a potential substitute of kerosene or diesel. So it can be used for running diesel engine, lighting in the rural houses, etc.

Table 2. Physical properties of Karanj seeds

Weight of seed, g	Av. length, L, mm	Av. width, W, mm	Thickness, mm	Slender ratio (L/W)	Bulk density, g/cc
1.19	17.88	14.42	6.26	1.24	0.56
(0.50)	(0.60)	(2.94)	(1.08)	(0.09)	(0.08)

N.B. Values inside the parentheses indicate SD.

5.9. Physical Properties

Some Physical properties such as physical dimensions, slender ratio, weight, bulk density of Karanj seeds at a moisture content range of 10.91% (d.b.) are detailed in Table 2.

Karanj seed was flattened with a circular shape having a slender ratio of 1.24. Jatropha seed was much larger than Castor seed although both the seeds were of similar shape with a slender ratio of 1.78 and 1.87 respectively having a cylindrical shape with rounded tips. Rubber seed was the largest among the four types of seeds under study. Its shape was rectangular with rounded tips and slightly flattened thickness having a slender ratio of 1.13. Bulk density of Karanj and Castor seed were found to be around 0.56 g/cc while Jatropha and Rubber seeds showed the lower bulk densities of 0.32 g/cc and 0.25 g/cc respectively. This is because both the latter mentioned seeds had hollow space between the endosperm and the outer hard shell. Moreover, endosperm of these two types of seeds was found to be spongy.

5.10. Mechanical Properties of the Seeds

Moisture content of the seeds was in the range of 9-14% (w.b.). Hardness of Karanj seeds was found to be around 1.8 kg. The reason was most part of the seeds was soft endosperm covered by a comparatively soft shell. Crushing strength of Karanj seeds were found to be 24 kg/cm^2 (Khan, 2008).

5.11. Oil Content and Cake Yield of the Seeds

The oil content of the Karanj seed was found to be 31.75% while the cake yield was 68.81%. These values are comparable with those of Jatropha, another potential source of biodiesel (Hussain *et al.*, 2008).

Conclusion

Karanj is an amazing tree which has high potential to withstand salinity and also able to grow on the river banks against strong current. So, these trees can be planted in the costal belts, islands for not only producing biofuel but also for protecting the costal areas and islands from cyclone, creating green forest which will promote tourism, protect soil from erosion and also for land reclamation from the sea bed. Kaanj seeds contain about 31.75% oil which shows its high potential as a source of biodiesel. Moreover, the cake (68.81%) can be a rich source of biofertilizer. Therefore, by planting Karanj in the hilly and plain lands specially in rural areas, poor people will be able to reduce their daily biomass requirements by replacing with biofuel derived from the Karanj seeds For this cause, Government must focus a long term national bio-fuel policy for the improvement of the social economical condition of the rural people and to promote all kinds of assistance for further research and extension of the energy trees plantation in the country.

REFERENCES

Anonymous. (2008a). Honge oil as a Biodiesel, Web: http://energy2.wordpress.com/2008/06/23/honge-oil-as-a-biodiesel/Last accessed: 10.09.08.

Anonymous. (2007b). '*Comparison of properties of karanj and diesel oil'*. Web: http://www.medorsbiotech.com/comparision1.htm Last accessed: 15.09.08.

Anonymous. (2007c). *Material, 'Quality Assurance'*, New Directions Aromatics Inc., Brampton on L7A 1C5, Revision Date: 8th May 2006. Web; http://www.newdirections-aromatics.com/ Last accessed: 15.09.08.

Antrim, A. (2008). '*Karanj for Wastelands'* by Appropedia. Web: http://www.appropedia.org /Pongamia Lat accessed: 15.09.08

Edward F. G. & Dennis G. W. (1994). '*Pongamia pinnata Pongam'*, Fact Sheet ST-498, a series of the Environmental Horticulture Department, Florida Cooperative Extension Service, Institute of Food and Agricultural Sciences, University of Florida. Publication date: October.

Fellows P. J. (1996). '*Food Processing Technology'*, Woodhead Publishing Limited, Cambridge, England, p146.

Hussain M. D., Islam M. P., Haque M. A. & Habib M. H. (2008). *Feasibility study on the production of Bio-diesel from Jatropha Carcus in Bangladesh*, Annual Progress Report: 1.07.07 to 30.06.08, Bangladesh Agricultural University Research System (BAURES).

Islam M. P, Hussain M. D. & Panigrahi S. (2008). 'Tropical trees as a potential rural energy source of Bangladesh', *International Journal of Environment and Sustainable Development (IJESD)*, Vol.X(X), Inderscience Enterprises Ltd, UK.

Kartha, A. R. S. & Sethi, A. S. (1957). 'Cold percolation method for rapid gravimetrically estimation of oil in small quantities of oil seed', *Indian J. Agric. Sci.,* 27(2):211-217.

Khan, F. (2008). '*A study on biodiesel production from selected indigenous seeds available in Bangladesh'*, A Project report published by Department of Farm Power and Machinery, Bangladesh Agricultural University.

Kureel, R. S., Singh C. B., Gupta A. K. & Pndey A. (2007). *Karanja: A potential source of bio-diesel.* National oilseeds and vegetable oils development Board, Ministry of Agriculture, Govt. of India, pp. 1-22. Web: http://www.novodboard.com Last accessed: 15.09.08.

Satish, L. (2007). *'Karanj (Pongamia pinnata)'*, Web: http:// www.svlele.com, Last accessed: 21.06.07.

Weiss, E. A. (1971). '*Oilseed processing. In Castor, Sesame and Safflower*', Ch.33, 761-793. First Ed. Barnes & Noble, Inc. New York.

In: Characterization and Development of Novel Materials
Editor: Rafiq Islam
ISBN: 978-1-61324-149-3

Chapter 13

EXPERIMENTAL STUDY ON BEESWAX USING WATERJET DRILLING

M. Enamul Hossain[1*], Chefi Ketata[2] and M. Rafiqul Islam[2]
[1]Department of Petroleum Engineering,
King Fahd University of Petroleum & Minerals
Dhahran 31261, Kingdom of Saudi Arabia
[2]Department of Civil and Resource Engineering,
Dalhousie University Halifax, Nova Scotia, Canada

ABSTRACT

Conventional rotary drilling has been the main technique used for drilling in the oil and gas industry. However, this method has showed its limitations regarding the depth of the wells drilled in addition to the use of toxic drilling fluids. This paper proposes an alternative method that is waterjet drilling. This drilling technique is environmentally friendly, fast, and feasible. Based on laboratory experiments, empirical models have been developed to describe the jet-rock interaction simulated by using beeswax samples. The resourceful and conclusive results show that the depth and rate of penetration increase with time, temperature, and thermal exposure.

Keywords: waterjet; drilling engineering; beeswax; penetration depth; penetration rate.

1. INTRODUCTION

Drilling is one of the oldest technologies in the world. At present, drilling technology has modernized well profiles and directions. High pressure waterjet (HPWJ) technique is now used for horizontal drilling. Horizontal wells are being drilled across the reservoir for exposing a relatively large reservoir area which is used as drainage or injection. Most of the

* Corresponding author: E-mail: meanmul@kfupm.edu.sa or dr.mehossain@gmail.com.

new wells are completed without cementing or liner. Due to the long opening, acidization is used to remove mud cake. Since, this process is very costly, the HPWJ process is increasingly used. HPWJ is a cost effective and simple to handle process in drilling technology. This technology has been the topic of many researchers due to its variety usages in the industry. As a result, recently, researchers are very interested on this tropic.

Lia et al. (2001) pointed out that the combined cutting effect of waterjet and polycrystalline diamond compact (PDC) is very effective in very hard rocks. They did experiments on waterjet and PDC for rate of penetration in hard rocks. Arangath et al. (2002) discussed the high hydraulic horsepower jetting tool which is used for scale removal. They have also shown waterjetting in horizontal well drilling. Dunn-Norman et al. (2002) discussed processes for sustainable recovery of heavy oil from ultra-shallow reservoirs, using low cost, innovative horizontal drilling and completion methods. They have argued the use of waterjet drilling over 15,000 ft (5,000 m) which is more competitive with the conventional rotary drilling system. They also concluded that waterjet drilling methods appear most favorable for drilling horizontal wells in ultra-shallow reservoirs.

Recently, Hossain et al. (2010) have briefly described the contributions of different researchers on waterjet drilling and the importance of this technology. They have developed some empirical relations using waterjet drilling on rate of penetration, depth of penetration (DOP) based on paraffin wax as a sample. They also found out the variation of depth, rate of penetration (ROP), temperature, flow rate and pressure with time. The variations of depth and ROP with temperature, side effect and thermal exposure time have also been studied. They did not study the effects of gap variation between drill bit tip and sample top surface. The present study investigated this effect.

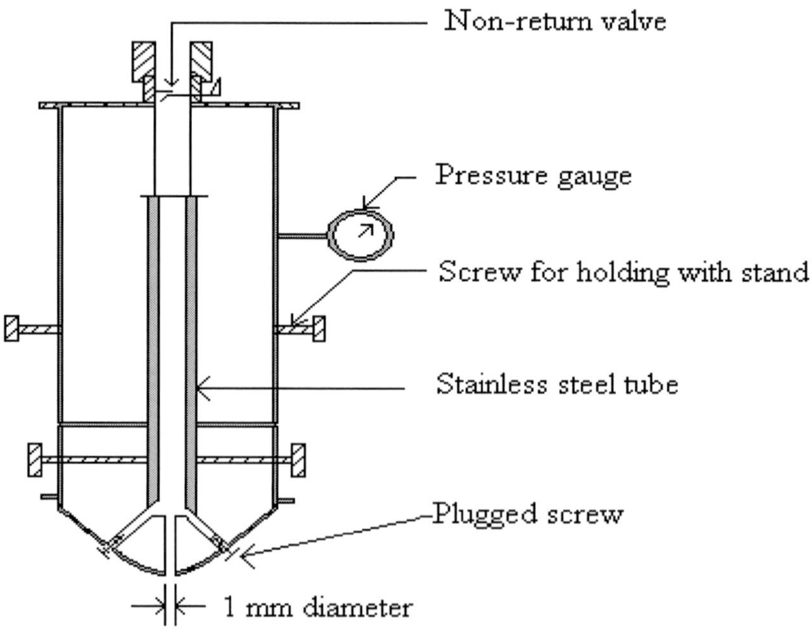

Figure 1. Stainless steel drill bit (redrawn from Hossain et al., 2006).

Figure 2. Experimental setup at the laboratory.

Figure 3. Beeswax sample for experiment.

To study the waterjet drilling system in more detail, beeswax sample has been used as a rock sample. The properties of beeswax samples are well described by Hossain et al. (2009a). Moreover, the factors that may affect the drilling activities are also explained by Hossain et al. (2009b). A series of laboratory tests were run to know the effect of water flow rate, the ROP, and pressure. This research examines the effect of jet time and temperature on the depth and ROP. In addition, the effect of change of waterjet pressure on the depth and rate of penetration is studied.

2. Experimental Setup

Figure 1 shows a laboratory experimental drill bit for waterjet drilling technique. Here normal tap water is used with a maximum of 72 psig pressure to create a waterjet through a 1-mm diameter hole at the tip of drill bit. A non-return valve has been used for protecting back flow and pressure. Two grooved screws are attached with the drill bit for holding it with a stand. A pressure gauge has been set with the bit for measuring the inside pressure of the bit.

Figure 2 shows a schematic view of the experimental setup used in the laboratory test. A garden hose was connected with the tape and drill bit top to get the continuous water flow. Figure 3 shows the beeswax that is used as a rock sample in the experiment.

3. RESULTS AND DISCUSSION

Figure 4 shows the dependence of different influential parameters such as DOP, ROP, flow rate, pressure of the drilling fluid system and temperature with time. The graph shows that depth is increasing with time almost linearly. However, after 20 minute, it does not increase any more. Initially, the rate of penetration decreases rapidly and with time, it decreases gradually. The flow rate is almost same during the experiment; however there is a little variation due to the variation of pressure in the laboratory water tap. It was attempted to keep the pressure constant during the experiment. However, there was little variation on pressure which shows in the figure (Figure 4). This was due to the direct use of tap water in the system. There is no temperature variation during the experiment in this case. So the trend is constant with time. The polynomial trendline for DOP with time is shown in Figure 5. The empirical relationship between these two parameters has been derived by best fit regression analysis. The relationship is represented by equation 1;

$$D = -0.024\,t^2 + 1.0302\,t + 0.2084 \text{ and } R^2 = 0.9959 \quad (1)$$

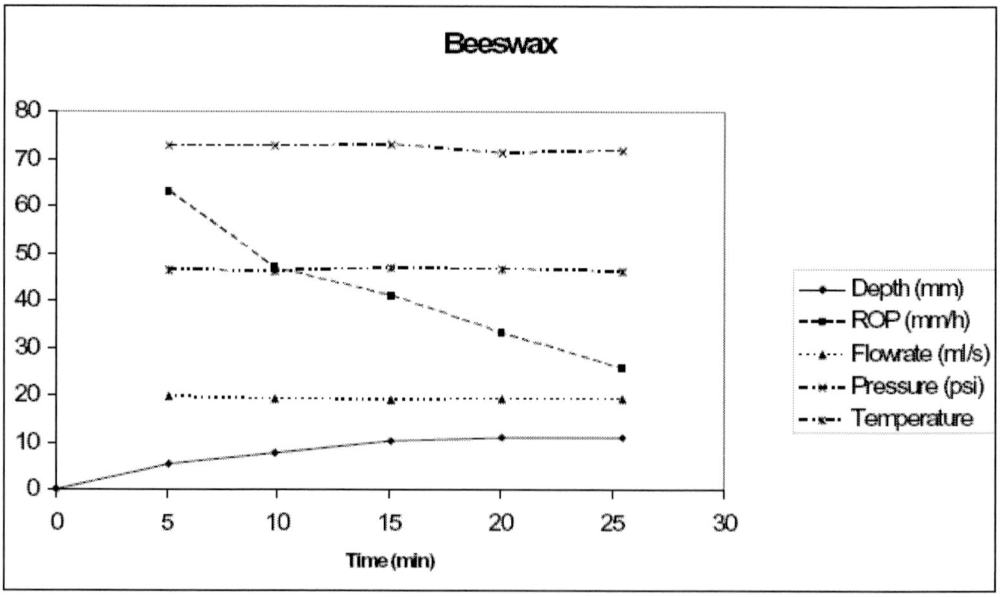

Figure 4. Dependence of different parameters on time.

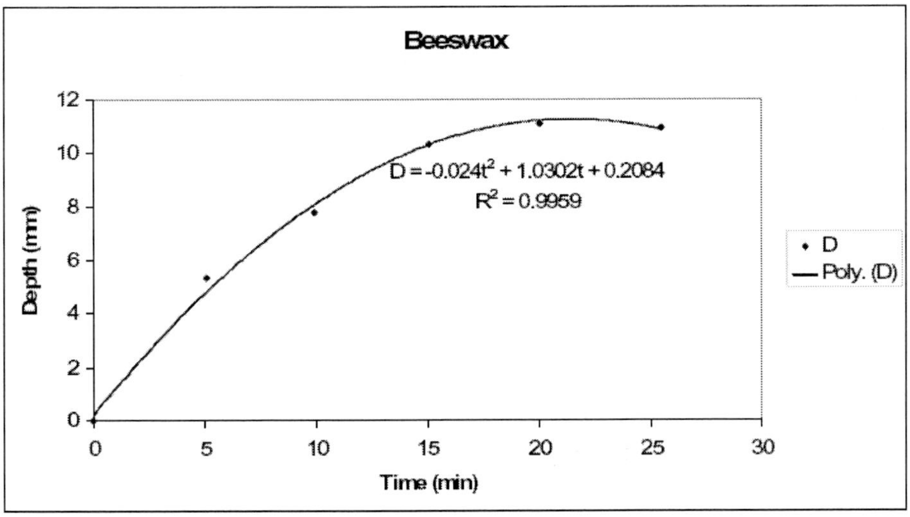

Figure 5. Dependence of depth on time for empirical relation.

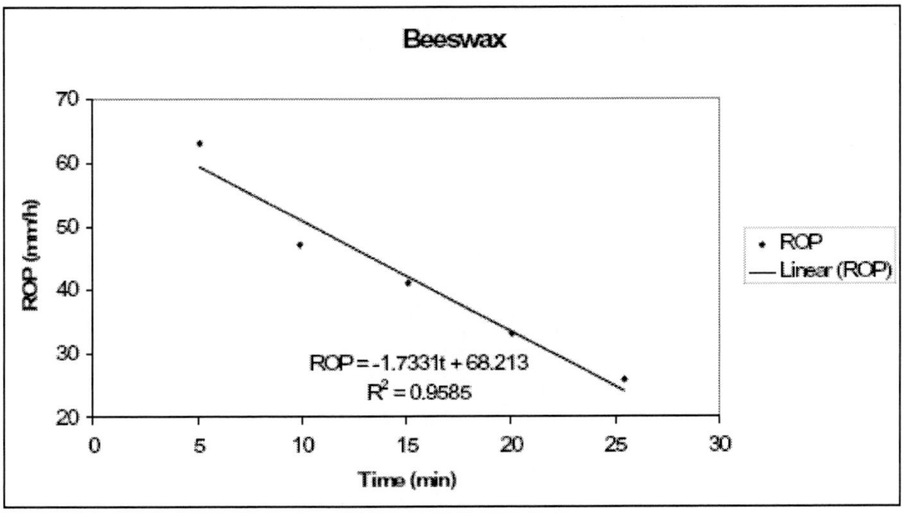

Figure 6. Dependence of rate of penetration on time for empirical relation.

The linear trendline for ROP with time is shown in Figure 6. The empirical relationship between these two parameters has been derived by best fit regression analysis which shows in equation 2.

$D_t = -1.7331\,t + 68.213$ and $R^2 = 0.9585$ (2)

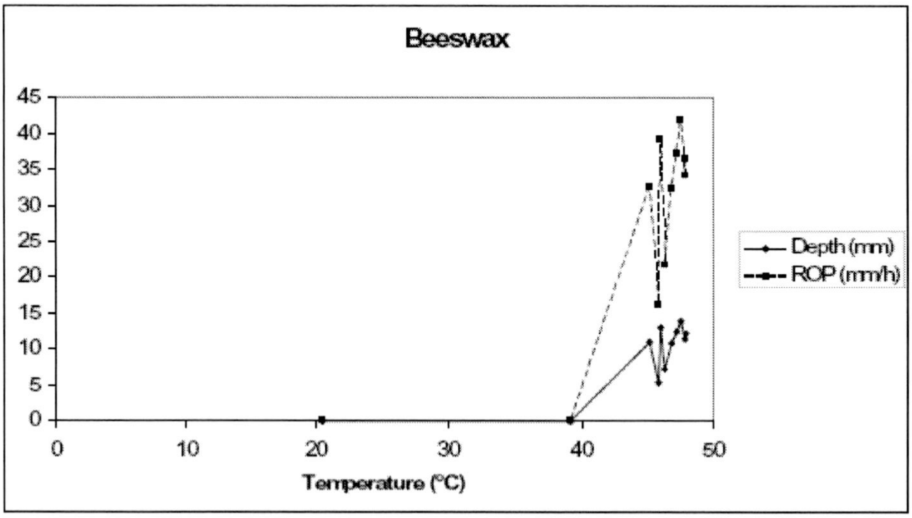

Figure 7. Effects of temperature on depth and rate of penetration.

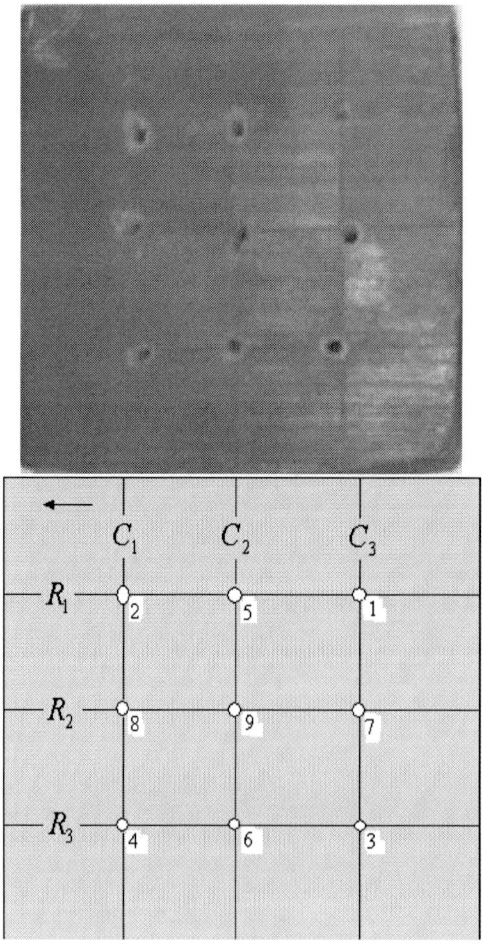

Figure 8. Systematic drilling to investigate side effects.

Figure 7 shows the affects the temperature of the drilling fluid on the DOP and ROP. There is no temperature affects up to 39.2^{0}C however after this temperature, the DOP and ROP increase with the increase of temperature. For both cases, the trend of variation is chaotic; there is no definite trend of variation for these two parameters. In addition, the depth and ROP are affected by other parameters such as side effect (Figure 8), thermal exposure, waterjet pressure and wax composition. To investigate the side effect, sets of experiment have been completed marking the sample in 2.3 cm interval from each other and each side (Figure 8). The figures show the orientation and DOP at different places within the sample. Each of holes has an interval of 2.3 cm between each others and side.

Figure 9 shows the column wise side effects on depth of penetration where distance from the centre of the wax sample has been considered as benchmark. Side effect is more effective in column 1 especially in reference point 2 and 8 (Figure 8). This is due to thermal effect on sample. Reference point 9 of column 2 is more affected than reference point 5 and 6. It seems that reference 9 should be less affected than point 5, 6 or 8 and 7 of column 1 and 3 respectively. This indicates that thermal effect is more sensitive than side effect. Reference point 7 shows more affected than reference point 1 in column 3. It is true that for beeswax, temperature is more influential parameters than side effects.

Figure 10 shows the row-wise side effect on DOP. Distance from the centre of the wax sample has been considered as the reference. Side effect is more effective in row 2 (Figure 8) due to thermal exposure effect. The behavior of row 3 is quite interesting where the trend of the line exhibits the side affect. Reference point 8 is the most affected point by side effect in the system.

Figure 11 shows the column-wise side effects on ROP. Distance from the centre of the wax sample has been considered as the reference. The trend of all the columns is same as explained for Figure 9. The row wise influence of side effects on ROP is represented by Figure 12. The behavior of the curves is almost same as DOP described for Figure 10.

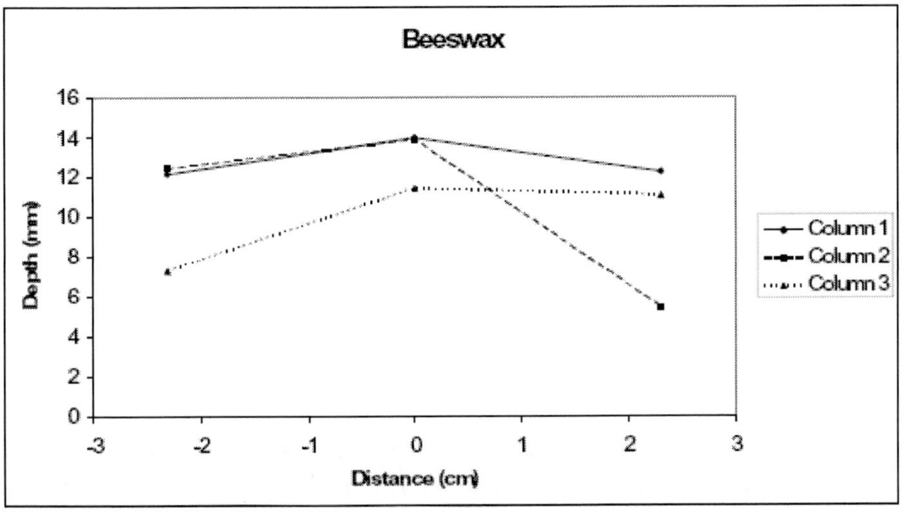

Figure 9. Side effects on depth of penetration (column-wise).

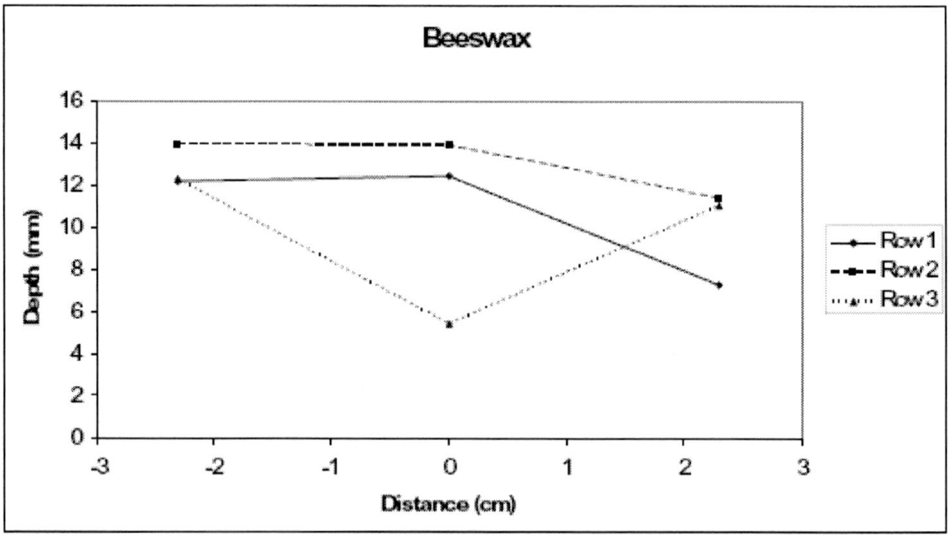

Figure 10. Side effects on depth of penetration (row-wise).

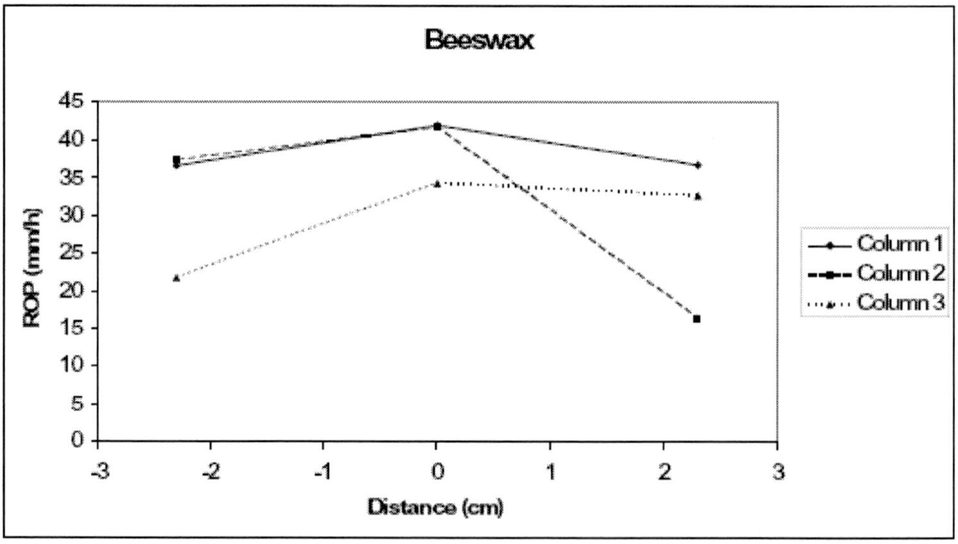

Figure 11. Side effects on rate of penetration (column-wise).

Figure 13 explains more clearly the effects of side on the beeswax sample. First point has been generated by averaging all the reference points of columns 1, 2, and 3 excluding the centre point 9 (Figure 8). The second point is the central point reference point 9. The same procedure is used to calculate both depth and ROP. The data trends depend on the drilling time, the water temperature, and the sample thermal exposure time. Figure 13 indicates the side effect for the sample experiments with varying depths and rates of penetration. To eliminate the sampling errors that affect these variables, every experiment should be conducted by an individual sample keeping sufficient distance from the edges i.e. at the center point.

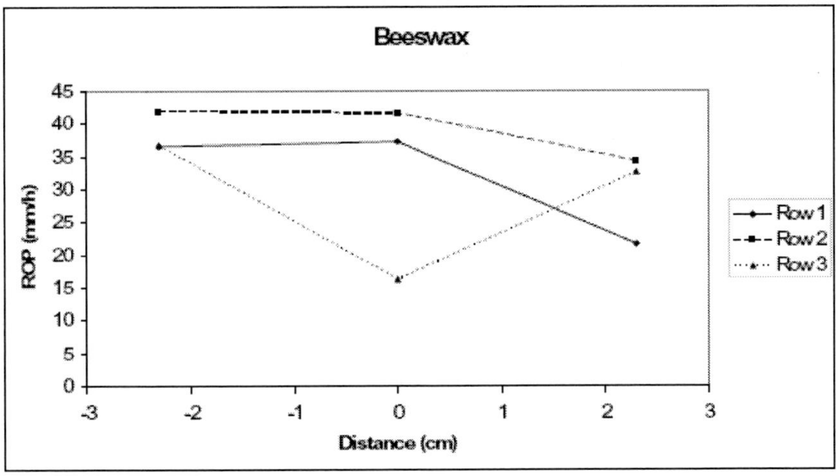

Figure 12. Side effects on rate of penetration (row-wise).

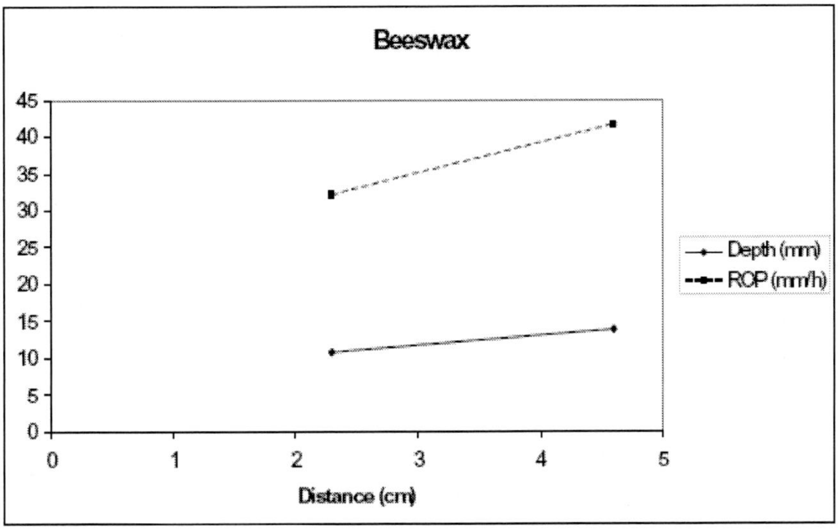

Figure 13. Side effects on depth and rate of penetration (towards centre from edge of the wax sample).

Thermal exposure time of the wax sample is investigated since the same wax sample was drilled at high temperatures and for many holes consecutively (Figure 14). As thermal exposure time increases, depth and ROP increases with time. It is obvious that there is a thermal effect during any hot water action. However, initially the influence is much more. The temperature is more sensitive in both depth and ROP. When the beeswax sample temperature reaches the drilling fluid temperature, it becomes steady. This indicates that thermal effects on the sample are limited by its drilling fluid temperature. If the temperature of the drilling fluid increases, thermal exposure time increases. The longer the thermal exposures time the more depth and ROP. Therefore, to soften the rock matrix, thermal action is helpful during waterjet drilling. As long as time progresses, thermal action plays a role in the rock matrix system shown in this figure.

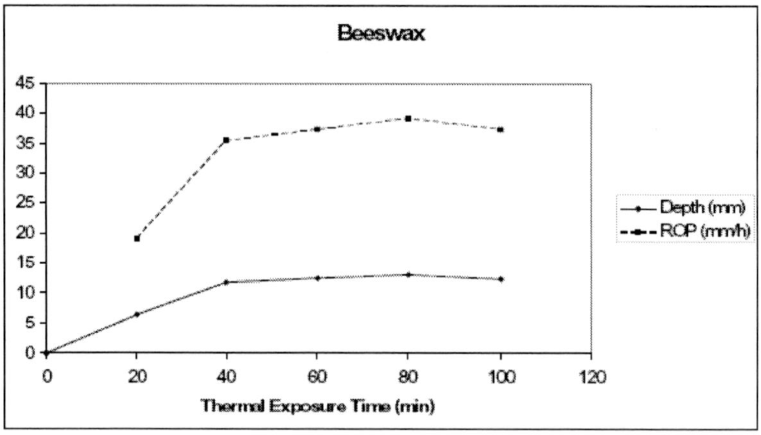

Figure 14. Variation of depth and rate of penetration with thermal exposure time.

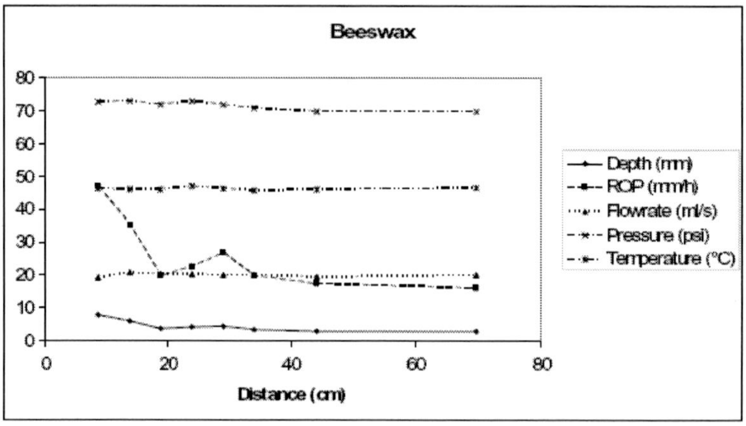

Figure 15. Dependence of different parameters on distance between jet nozzle and beeswax surface.

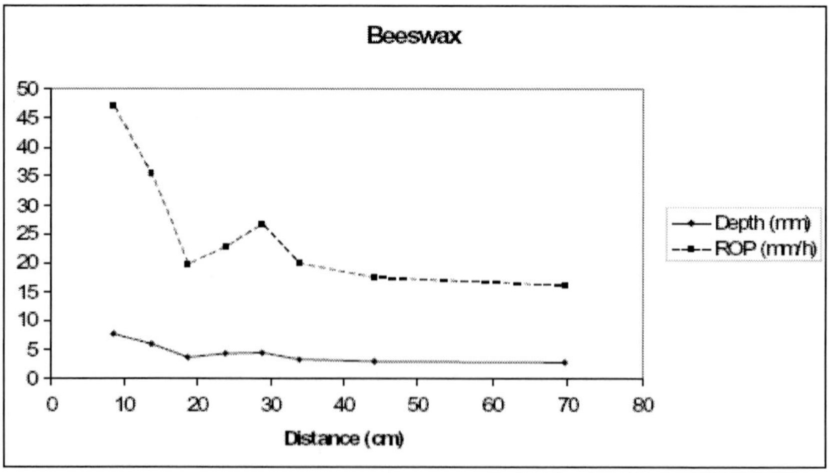

Figure 16. Dependence of depth and ROP on distance between jet nozzle and beeswax surface.

Figure 15 shows the affects of different influential parameters such as DOP, ROP, flow rate, pressure of the drilling fluid system and temperature with distance between jet nozzle and beeswax surface. The graph shows that depth is decreasing with distance between jet nozzle and beeswax surface. However, after 44 cm, it is almost linear. Initially, the rate of penetration decreases rapidly with distance between jet nozzle and beeswax surface up to 18.8 cm and then there is a fluctuation with distance up to 34.0 cm. After this distance, it decreases gradually. The flow rate is almost same during the experiment; however there is a little variation due to the variation of pressure in the laboratory water tap. It was attempted to keep the pressure constant during the experiment. However, there was little variation on pressure which shows in the figure (Figure 15). This was due to the direct use of tap water in the system. There is also some variation of temperature during the experiment which is shown in the figure.

Figure 16 shows the affects of important parameters – depth of penetration, and ROP with distance between jet nozzle and beeswax surface. The trend is more clearly visible in this figure.

CONCLUSION

The conventional practice in the oil industry is to use different drilling techniques where huge capital is involved. The technology is also more complicated to handle. In this regard, waterjet drilling is simpler and needs less capital involvement. Experiments show that DOP and ROP depend on pressure, temperature and flow rate of the waterjet. The ROP decreases with time, however DOP increases with time. The variation of depth and ROP is established with thermal exposure time. There is a negligible side effect during the experiment. It decreases from the edge towards the centre of the sample. The variation of depth and ROP on distance between jet nozzle and beeswax surface are also studied and shown. Empirical models are developed to describe the depth and rate of penetration with time.

NOMENCLATURE

A	=	cross sectional area of drill bit tip, mm^2, $[L^2]$
D	=	depth of penetration, mm, $[L]$
D_t	=	rate of penetration (ROP), mm/hr, $[L/t]$
p	=	pressure of the system, psi, $[M/Lt^2]$
Q	$= A \times u =$	flow rate, cc/sec, $[L^3/t]$
T	=	temperature, $[T]$,°C
t	=	time, minute, $[t]$
t_T	=	thermal exposure time, minute, $[t]$
u	=	waterjet velocity at the tip of the drill bit, cm/sec, $[L/t]$

ACKNOWLEDGMENT

The authors would like to thank the Atlantic Canada Opportunities Agency (ACOA) for funding this project under the Atlantic Innovation Fund (AIF).

REFERENCES

Arangath, R., Ogoke, V. C. & Onwusiri, H. N. (2002) 'A Cost-Effective Approach to Improve Performance of Horizontal Wells Drilled in High Permeability Formations', SPE 73786, presented at the *SPE International Symposium and Exhibition on Formation Damage Control* held in Lafayette, Louisiana, 20–21 February.

Dunn-Norman, S., Gupta, A., Summers, D. A., Koederitz, L. F. & Numbere, D. T. (2002) 'Recovery Methods for Heavy Oil in Ultra-Shallow Reservoirs', SPE 76710, presented at the *SPE Western Regional/AAPG Pacific Section Joint Meeting* held in Anchorage, Alaska, U.S.A., 20–22 May.

Hossain, M. E., Ketata, C. & Islam, M. R. (2010). 'Scaled model experiments of waterjet drilling', *Advances in Sustainable Petroleum Engineering Science*, 2(1), in press.

Hossain, M. E., Rahman, M. S., Ketata, C. & Islam, M. R. (2009d). 'SEM-based structural and chemical analysis of paraffin wax and beeswax for petroleum applications', *Journal of Characterization and Development of Novel Materials*, 1(1), 21 – 38

Hossain, M. E., Ketata, C. & Islam, M. R. (2009b). 'A scaled model study of waterjet drilling", *Petroleum Science and Technology*, 27, 1261 – 1273.

Lia, X. B., Summersb, D. A., Rupertb, G. & Santib, P. (2001) 'Experimental investigation on the breakage of hard rock by the PDC cutters with combined action modes', *Tunneling and Underground Space Technology*, Vol. 16, pp. 107-114.

In: Characterization and Development of Novel Materials
Editor: Rafiq Islam
ISBN: 978-1-61324-149-3

Chapter 14

SUPPORTED IONIC LIQUID MEMBRANES: PREPARATION, STABILITY AND APPLICATIONS

A. P. de los Ríos*[*]*, F. J. Hernández, L. J. Lozano and C. Godínez
Department of Chemical and Environmental Engineering,
Technical University of Cartagena, Campus La Muralla,
Cartagena, Murcia, Spain

ABSTRACT

The use of supported ionic liquid membranes, which consist of a thin microporous support whose pores are filled with a liquid, have been extensively studied during the last decade. Supported ionic liquid membranes have being recognized as an effective method for the selective separation of organic compounds and mixed gases. The unique properties of ionic liquids such as their negligible vapour pressure, their greater viscosity and the possibility of minimizing their solubility in the surrounding phases by adequate selection of the cation and anion could overcome the instability associated with supported liquid membranes based on organic solvents. Furthermore, the possibility of design suitable ionic liquids for specific separation problems has also opened up new fields of application of supported ionic liquid membranes. In this review an overview is given of recent advances in supported membranes based on ionic liquids, including issues such as methods of preparation, physical–chemical characterization, stability and long–term performance and fields of application of supported ionic liquid membranes.

1. INTRODUCTION

New processes based on supported ionic liquid membranes (SILMs) have been proposed for the highly selective separation of organic compounds involved in the synthesis of pharmaceutical and fine chemicals, such as alcohol, esters, organic acids [1] and amino acids [2] in the last decade. This new materials could overcome the relative inestability associated with supported liquid membranes based on organic solvents [3]. The latter membranes

[*] Corresponding author: E–mail address: antonia.perez@upct.es.

usually experiment a progressive loss of the liquid embedded in the membrane into the surrounding due to liquid vaporization, dissolution into the adjacent liquid phases and/or shifting from the porous structure of the supported membrane under low–pressure gradients [4]. The use of ionic liquids could overcome these inconveniences due to their inherent characteristic properties [5–7]. Ionic liquids (ILs) are organic salts which are liquids at room temperature consisting in an organic cation, the most commonly used are dialkylimidazolium and tetraalkylammonium, and a polyatomic inorganic anion (e.g. BF_4^-, PF_6^-, NTf_2^-) [8]. The most important advantage of ILs is their non–detectable vapour pressure, which make them environmentally benign solvents compared with volatile organic solvents. The use of these new solvents as a liquid membrane phase results in the stabilization of the SLMs due to their negligible vapour pressure, the possibility of minimizing their solubility in the surrounding phases by adequate selection of the cation and anion, and their high viscosity, which could reduce displacement of the liquids from the micron pores under pressure. [9]. Ionic liquid viscosities at room temperature range from a low of around 10 cP [10] to values in excess of 500 cP [11]. For comparative purposes, the viscosities of water, ethylene glycol, and glycerol at room temperature are 0.890, 16.1, and 934 cP, respectively [12]. Other interesting properties of ILs in the context of liquid membranes are their high chemical and thermal stability, high ion conductivity [13] and high solvent power [14]. All these mentioned properties have led ILs to being considered "green designer solvents" [15]. Indeed, ionic liquids has received growing attention during the last decade as clean alternatives to classic organic solvents for use in a wide variety of chemical processes, such as separation and purification [1,16–20], reaction media [21–24] and chemical catalysis [25–27].

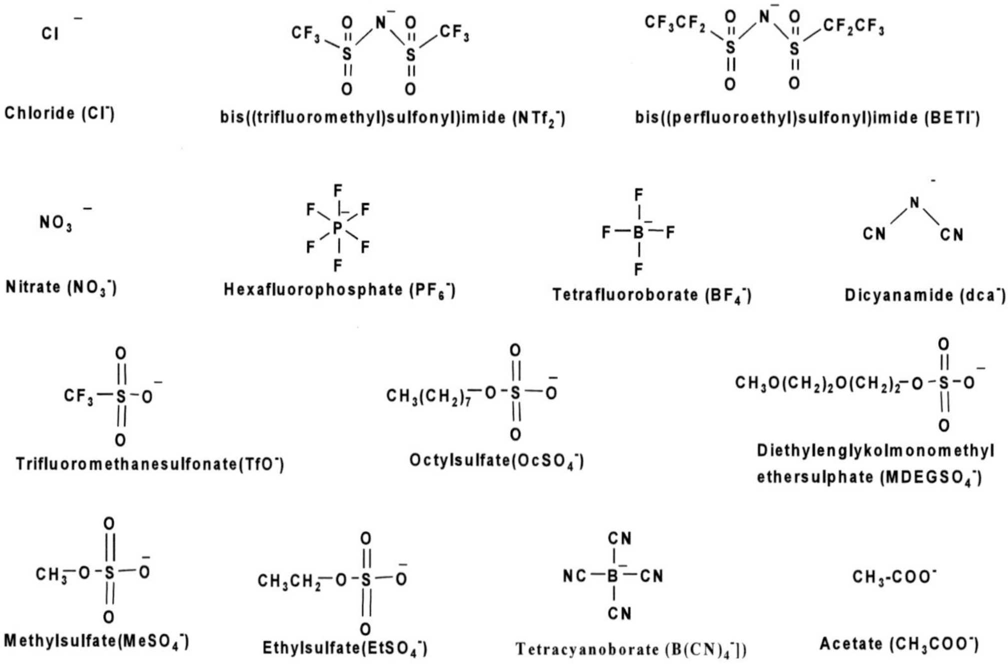

Figure 1. Example of ions involved in ILs.

An overview of the recent advances in supported ionic liquid membrane technology is given in this work, including issues such as methods of preparation, physico–chemical characterization, stability and fields of application of supported ionic liquid membranes.

2. Methods of Preparation and Characterizaton of Supported Ionic Liquid Membranes

Due to the relatively high viscosity of the ionic liquids, the method of preparation of supported ionic liquid membranes can play an important role on the membrane performance. Three methods are commonly used for the preparation of supported ionic liquid membranes: direct immersion, pressure or vacuum [2,19,28]. In the former case, immobilization takes places by contacting the supporting membrane with the ionic liquid, allowing it to soak up the liquid [19]. In the pressure method, the immobilization is achieved by placing the membrane in an ultrafiltration unit, adding an amount of ionic liquid and applying nitrogen pressure to force the ionic liquid to flow into the pores of the membrane [2]. In the vacuum method, the supporting membrane is submerged in a volume of IL and vacuum is applied to release all air occluded in the pores of the membranes [28]. After all procedures, the excess ionic liquid should be removed from the membrane surface either by leaving to drip overnight or by blotting with paper tissue. All these procedures are based on imbibition of the ionic liquid on the polymeric membrane. Due to the different viscosity of the ionic liquids and taking into account that imbibition is a time a dynamic process slowed down with viscosity, equal times in membrane preparation.are required to compare the behaviour of the resulting supported ionic liquid membranes.

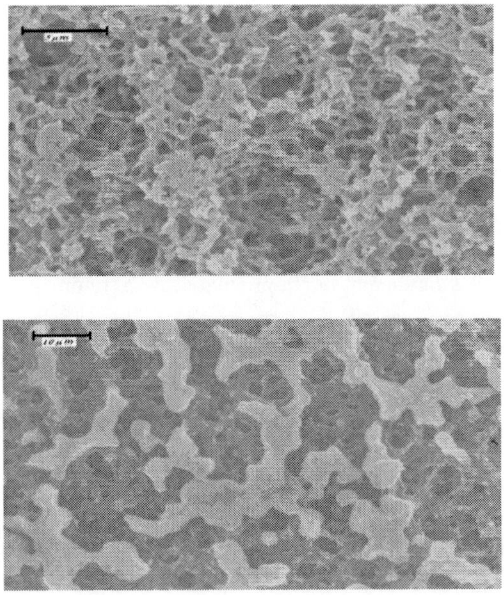

Figure 2. Scanning electron micrographs of the (a) Nylon membrane (3040x) (scale bar = 5 μm) and the Nylon membrane impregnated with $[bmim^+][PF_6^-]$ by pressure (b) (1010x) (scale bar = 10 μm).

Scanning electron–microscopy (SEM) combined with Energy Dispersive X–ray (EDX) analysis could be used to physico–chemical characterization of SILMs [28]. This technique allows the characterization of the membrane surface morphology and the examination the global chemical composition of the membranes and the distribution of the ILs within them. Figure 2 shows examples of SEM micrographs of a plain Nylon membrane and supported liquid membranes based on [$bmim^+$][PF_6^-] prepared by using the pressure method [28].

SEM micrograph of the Nylon membrane shows a highly porous material consisting of macropores of about 1 μm average size. The morphological study of SILM based on [$bmim^+$][PF_6^-] show that, after impregnation by pressure method, the IL is homogeneously distributed in the pores of the membrane, except for the largest macropores that remained partially filled. These pictures also show the presence of a small amount of excess IL located on the external membrane surface.

The EDX spectrum of the Nylon membrane (not shown) present the characteristic peaks assigned to the C, O and N Kα lines, the last two as a single broad peak unresolved due to their proximity and small concentration. The presence of these chemical elements corresponds to the chemical formulation of polyamide [28].

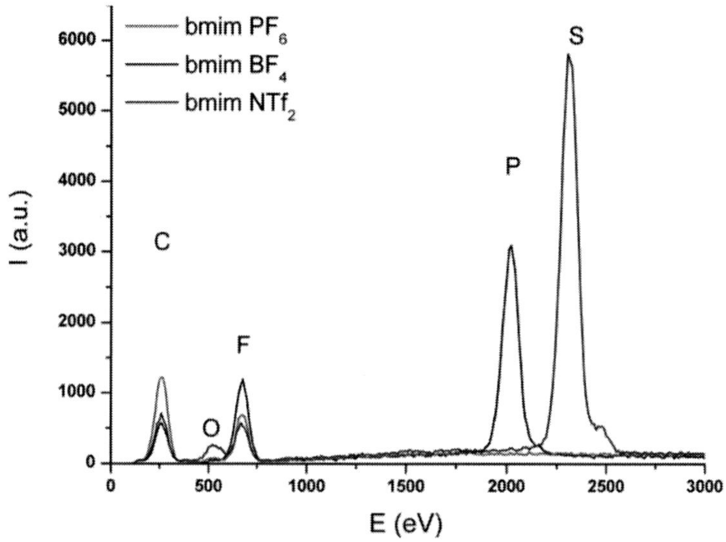

Figure 3. EDX spectra of the Nylon membrane impregnated with [$bmim^+$][PF_6^-], [$bmim^+$][BF_4^-] and [$bmim^+$][NTf_2^-].

The EDX study of supported ionic liquid membranes is based on the selection of characteristic elements of interest in each ionic liquid, for example, F and P for [$bmim^+$][PF_6^-], F for [$bmim^+$][BF_4^-], and F and S for [$bmim^+$][NTf_2^-] as shown in Figure 3. The relative peak heights of identical elements in the different compounds are grossly related with their respective concentrations [28].

It has been also demostrated by SEM–EDX and weight studies that the amount of IL immobilized is independent of the ionic liquid used when the immobilization is carried out under pressure [28]. In contrast, following immobilization under vacuum, less ionic liquid is absorbed into the membranes when more viscous ILs are used. This behaviour could be explained by the fact that the high viscosity of these ILs makes it difficult their penetration into the middle of the deeper pores of the membrane under vacuum. From these findings it was concluded that immobilization under vacuum can be considered suitable for low viscosity ionic liquids, with the added advantage that it is also an easier preparation method. However, for high viscosity ionic liquids, immobilization of the ionic liquid should use pressure in order to ensure that all membrane pores are filled with liquid.

The nature of the supporting membrane also plays an important role in the performance of supporting ionic liquid membranes. In this context, *De los Rios et al.* [5] studied the used of two polymeric membranes, Nylon and Mitex, as supporting membranes. Nylon membrane was a hydrophilic polyamide membrane with a pore size of 0.45μm and a thickness of 170 μm. Mitex membrane was a hydrophobic polytetrafluoroethylene membrane with a pore size of 10 μm and a thickness of 130 μm. It was observed that less ionic liquid was absorbed into the Mitex membranes, which was explained by the different textural properties and the high hydrophobic character of these membranes, which probably restrict interaction with the hydrophilic ionic liquids used [29].

The use of nanofiltration membranes as supporting membranes have been also reported [30]. In this case, direct filtration of ionic liquids through the nanofiltration membrane was not possible at a gas pressure up to 7 bars. The ionic liquids make up of cations which contain straight or branched hydrocarbon chains were easily absorbed into the polymeric membrane allowing the nanoporous structure soaked up and saturated with the ionic liquids.

3. Stability of Supported Ionic Liquid Membranes

One of the most important properties for the industrial application of SILMs is their stability. Therefore, stability studies of SILMs prepared by several methods at different operating conditions (e.g. surrounding phases, pressure and temperature) have been reported [5,28,31–38].

The immobilization method was also found to have influence on the membrane stability. A comparative study of the preparation of SILMs by two different methods, under pressure and vacuum were reported by Hernández–Fernández *et al.* [28]. They use the ionic liquids $[bmim^+][Cl^-]$, $[bmim^+][BF_4^-]$, $[bmim^+][PF_6^-]$ and $[bmim^+][NTf_2^-]$ as liquid phase supported on a Nylon membrane. Small losses of ionic liquid were observed when the ionic liquid was immobilized under pressure after 7 days of operation in a diffusion cell using n-hexane/n-hexane as surrounding phases. However, the losses of IL were higher when immobilization was carried out under vacuum, specially with the most viscous ionic liquids ([bmim+][PF6−] and [bmim+][Cl−]). This behaviour was explained by the fact that the higher viscous ILs makes difficult their penetration into the middle of the deeper pores of the membrane and therefore, the ionic liquid was mainly immobilized on the most external layer of the membrane and, consequently, the immobilized ionic liquid is more easily removed during operation.

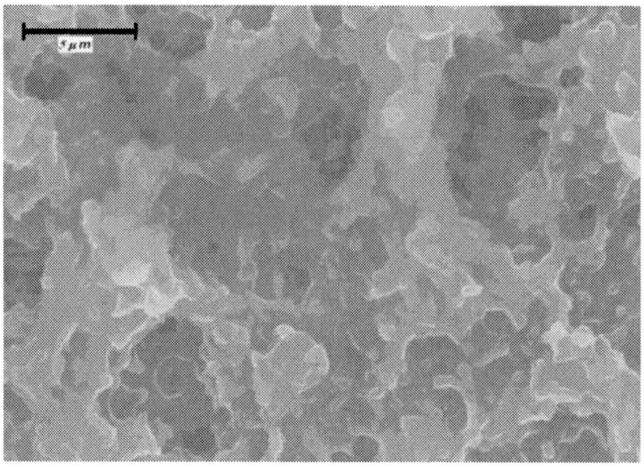

Figure 4. Scanning electron micrograph of the Nylon membrane impregnated with [bmim$^+$][PF_6^-] after seven days´ operation (3040x) (scale bar = 5 μm) in n–hexane/n–hexane.

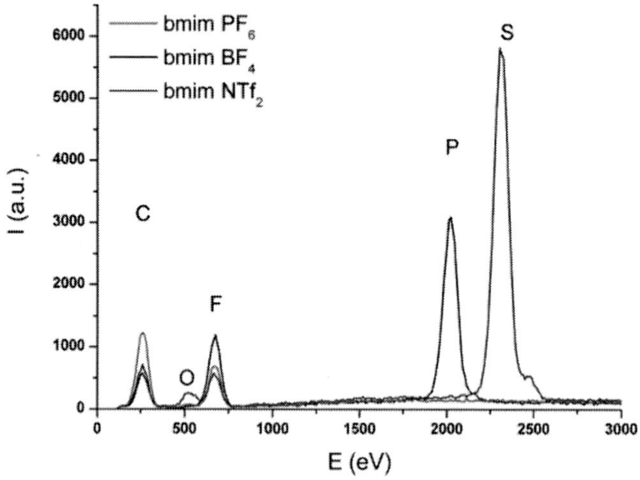

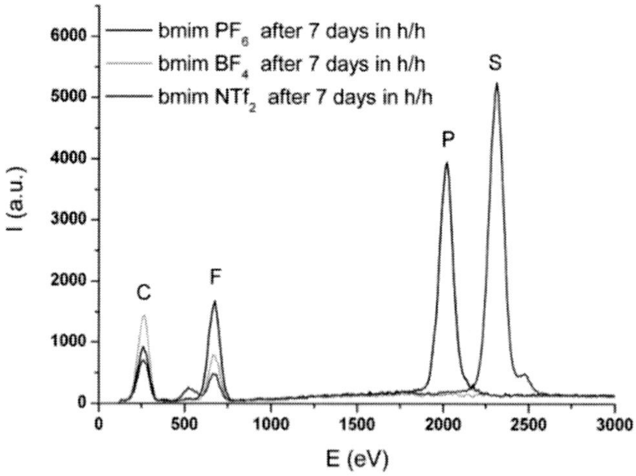

Figure 5. EDX spectra of the Nylon membranes impregnated with [bmim$^+$][PF_6^-], [bmim$^+$][BF_4^-] and [bmim$^+$][NTf_2^-] (a) before and (b) after seven days´ operation in n–hexane/ n–hexane.

The influence of the ionic liquid composition on the stability of the resulting SILM in organic environments has also been analyzed. In this context, de los Rios. *et al.* [5] analyzed the stability of SILMs based on $[bmim^+][PF_6^-]$, $[bmim^+][BF_4^-]$ and $[bmim^+][NTf_2^-]$ using Nylon membranes as support in n–hexane/n–hexane. Stability tests were performed by keeping the respective impregnated membranes immersed for a week in a diffusion cell including two independent compartments and using n–hexane/n–hexane as the respective feed and receiving phases. Figure 4 shows the microstructure of the Nylon membrane impregnated with $[bmim^+][PF_6^-]$ after cell operation, which could be compared with the appearance of the SILM before cell operation (Figure 2b).

It was concluded that most of the ionic liquid from the external surface of the membrane disappeared during cell operation. However, comparison of the SEM–EDX spectra taken from membranes before (Figure 5a) and after (Figure 5b) immersion in the n–hexane/ n–hexane solution showed that they were very similar. The EDX spectra are taken from a sample of up to a few micrometers thick/depth and shows that the contribution of ionic liquid within the membrane pores is more important than the accumulated liquid found on the surface. Consequently, from the SEM study it was deduced that only the ionic liquid deposited on the external membrane surface has been stripped off during operation. The amount of ionic liquid retained in the membrane pores, however, was apparently kept constant and, consequently, the membrane was stable against the possible solvent action of n–hexane.

From the SEM–EDX study, it was concluded that, in all three cases studied of ILs supported on Nylon membranes, no substantial losses of ionic liquid occurred after a week of operation in an n–hexane/n–hexane diffusion cell. Only minimal losses of the ionic liquid accumulated on the external surface of SLMs were detected qualitatively by the SEM pictures.

The stability of a SILM based on $[bmim^+][BF_4^-]$ supported in a Nylon membrane has been also analyzed in other organic solvents, such as *n*–hexane/*tert*–butyl methyl ether and *n*–hexane/dimethyl sulfoxide [31]. The SEM–EDX study of the membranes after continuous operation showed that the stability of the supported liquid membrane increase with the decrease of the polarity of the solvent used. The relationship established between polarity and stability allows the design of more stable SILMs for use in organic environments.

The stability of SILMs based on 1–n–alkyl–3–methylimidazolium hexafluorophosphate (n=4,8,10) supported in PVDF membrane towards contacting aqueous phase has been also analysed by several authors [2,6,32]. The resulting SILMs were stable at the assayed conditions. These authors highlighted the importance of the consideration of two main possible effects on the performance and stability of SILMs in water mediums: (a) the loss of ionic liquids organic phase from the supporting membrane to the adjacent aqueous solutions by dissolution/emulsification; and (b) the formation of water microenvironments inside the ionic liquids organic phase, which constitute new, non–selective environments for solute transport, leading to a deterioration of the SLM performance and selectivity.

Stability studies of SILM have been also carried with gases as contacting phases [33-37]. The stability of SILMs based on the ionic liquid $[Et(OEt)_2^+][PF_6^-]$ supported in Durapore hydrophobic PVDF was studied at 25°C with a pressure difference of 1.2 bars [33]. Experiments concerning stability showed that after eight cycles the permeability of H_2 slightly decreased while the CO_2 /H_2 ideal selectivity improved. Other interesting example was reported by Hanioka *et al.* [34], who examined the long term stability of a SLM based on a functionalized ionic liquid ($[NH_2pmim^+][NTf_2^-]$) in the separation of the CO_2/CH_4 gas mixture at atmospherical pressure. These authors did not found observable change in permeability or selectivity after 260 days operation. Scovazzo et al. [35] analyzed the stability of SILMs used in the separation of gas pairs CO2/CH4 and CO2/N2 using continuous flows of the mixed gases. They achieved long–term stability in continuous operation, from 24 to 106 days, by using imidazolium ionic liquids supported in hydrophilic–PVDF or hydrophilic–PES membranes, respectively, without performance degradation under CO2–partial pressures of at least 2.07 bars.

Temperature stability has a great interest for some applications of SILMs in gas separation, such as capture of CO2 from coal gasification plants. Ilcovich *et al.* [36] analyzed the stability of a SILM based on $[hmim^+][NTf_2^-]$ supported in a polysulfone organic membranes in the selective separation of CO_2 from He at high temperature. This membrane was found to be stable upto 125 °C, the failure of the membranes above that temperature being attributable to support failure rather than any effect on the ionic liquid. Recently, Myers *et al.* [37] reported operating $[hmim^+][NTf_2^-]$ supported in Nylon membranes up to 300◦C. It was also found that permeability in this $[hmim^+][NTf_2^-]$–membrane increased with temperature while the selectivity decreased.

SILMs are usually prepared using microfiltration membranes an operated with low varying pressure. To avoid the pitfall of liquid instability associated with micron porous membranes, nanofiltration membranes were used which greatly reduce the instability problem, but potentially at the expense of increased gas transport resistance provided by the NF membranes [27]. Experimental stability tests demonstrated that the impregnated ionic liquids did not discharge from the NF membrane structure even under a high transmembrane pressure of 10 bars. The test results suggest that SILMs supported on nanofiltration membranes offer a good stability able to sustain high gas phase pressure relevant to practical applications.

4. Fields of Application of Supported Liquid Membranes

The use of supported ionic liquid membranes in different fields of application have received growing attention in the last decade, due to the advantages that SILMs offer, presenting the number of paper published in this field per years an exponential growth. One of the most studied applications of SILMs is the selective separation of organic compounds. The first example was reported by Branco et al. [13], who studied the selective transport of 1,4–dioxane, 1–propanol, 1–butanol, cyclohexanol, cyclohexanone, morpholine and methylmorpholine as a model seven–component mixture of representative organic compounds. For that, four ionic liquid based on the 1–n–alkyl–3 methylimidazolium combined with the anions hexafluorophosphate or tetrafluoroborate, immobilized in different organic polymeric membranes were used. The use of the ionic liquid $[bmim^+][PF_6^-]$ immobilized in a polyvinylipolyvinylidene fluoride membrane, allowed an extremely highly selective transport of secondary amines over tertiary amines (up to a 55:1 ratio).

Other interesting examples were reported on the use of SILM for the selective separation of the substrates and products of transesterification reactions. The biosynthesis of organic esters commonly used in the perfumery and flavour industries can be carried out by transesterification from vinyl esters and alcohols catalysed by enzymes in non–conventional media (i.e n–hexane [38] and ionic liquid [39]) at low water content. When the biochemical reaction reach the equilibrium, the reaction medium could consist of a mixture of alcohol, vinyl ester, organic acid and alkyl ester (flavour ester). The possibility of using SILMs for the selective separation of these reaction mixtures has been extensively analysed [38,17,40–42]. In these cases substantial permeability differences between the target compounds were found on the basis of not only their functional groups but also of their alkyl chain lengths. It was also found that selectivity values were mostly dependent on the anion composition of the ionic liquid, being increasing for a given cation ([bmim]) in the sequence: $[OcSO_4^-] < [PF_6^-] < [NO_3^-] < [BF_4^-] < [dca^-] < [Cl^-] < [EtSO_4^-] < [MDEGSO_4^-] < [MeSO_4^-]$ [42]. These results were quite encouraging and suggested that these SLMs based on ILs could be use for the selective separation of the organic esters from the reaction mixture. SILMs have also be used for the separation of aromatic hydrocarbons from aliphatic hydrocarbons. In this context, the selective separation of benzene, toluene and *p*–xylene from n–heptane was analyzed using SILMs based on $[bmim^+][PF_6^-]$, $[hmim^+][PF_6^-]$, $[omim^+][PF_6^-]$ and $[Et_2MeMoEtN^+][Tf_2N^-]$ supported in a polyvinylidene fluoride membrane [14]. It was found that aromatic hydrocarbons were successfully transported through the membrane based on these ionic liquids, the maximum selectivity to n–heptane being reached using benzene in the aromatic permeation and $[bmim^+][PF_6^-]$ in the liquid membrane phase.

Other interesting field of application of supported ionic liquid membranes is the separation of mixed gases. Since SILMs may have the potential for industrial applications, specifically, low pressure systems such as the treatment of bio–methane from anaerobic digesters and CO_2 capture from flue gases, much effort and many resources have been expended on developing new SILMs [35,43-45]. In this context, Scovazzo *et al.* [35] explored the selective separation of the gas pairs CO_2/CH_4 and CO_2/N_2 using continuous flows of the mixed gases using $[emim][BF_4]$ $[emim^+][TfO^-]$ and $[emim^+][dca^-]$ supported in polyvinylidene fluoride and $[emim^+][Tf_2N^-]$, $[hmim^+][Tf_2N^-]$ and $[emim^+][BETI^-]$ supported in a polyethersulfone membrane. The highest CO_2/ CH_4 and CO_2/N_2 selectivities were 27 and 21.2 using the ionic liquids $[emim^+][BF_4^-]$ and $[emim^+][Tf_2N^-]$, respectively. These authors also pointed that for CO2-separations, there are two critical IL properties that effect SILM performance: IL molar volume and IL viscosity [43]. The selectivity is a function of IL molar volume while the CO2-permeability is a function of viscosity. To increase the selectivity of the process, Scovazo et al. [43] proposed using anions or cations that produce smaller molar volume ILs, since as it is predicted by The Camper Model's [46], selectivity increase exponentially as the IL molar volume decrease.

Supported ionic liquid membranes have also been successfully applied to the separation of various liquids or vapour mixtures by pervaporation and vapor permeation [47–49]. The major problem associated with solute recovery by these techniques is the restrictive compromise between selectivity and flux: High solute selectivity involves the use of conditions that lead to relatively low mass fluxes in the membrane. This problem would be overcome, if the feed solvent could not permeate through the membrane, for example using SLMs based on ILs. In this case, solute to solvent selectivity could be significantly increased and the process could be operated under conditions that provide high fluxes [47]. In this context, the use of SLMs based on the ionic liquid [Pr4N+][B(CN)4–] in a nanofiltration ceramic module allowed the removal of 1,3–propanediol from aqueous solution by pervaporation, increasing the selectivity of the process by more than two orders of magnitude while the permeability was only in one order of magnitude slower [47].

Supported ionic liquid membranes have also found application in analytical chemistry, specifically in the trace determination of toxic. Several examples of the application of ionic liquids as liquid membrane in hollow fiber supported liquid–phase microextraction have been reported [50,51]. This technique, combined with HPLC, was used for the determination of chlorophenols and sulfonamides in environmental water samples.

CONCLUSION

Supported ionic liquid membranes (SILMs) have emerged as novel material for a wide range of separation processes including organic molecules separation, gas separation and pervaporation process and analytical chemistry due to the possibility of design suitable *ionic liquids* for *specific* separation problems. Three methods are commonly used for the preparation of supported ionic liquid membranes: direct immersion, pressure or vacuum. Although, immersion and vaccum method are easier to applied, pressure method allows to obtain SILM with improved properties. Stability studies on SILMs have corroborated their high operational and structural stability *which makes possible their industrial application.* Another notable aspect is the possibility of tailoring their physicochemical properties of the SILMs by selection of the ionic liquids constituent of the liquid membrane which will open up new field of application of these SILMs. Futhermore, the use of ionic liquid instead of volatile organic solvents as liquid make SILM more environmentally benign materials. The latter mentioned characterics allow SILMs to being considered "green designer materials". The preliminary studies about supported ionic liquids membranes are quite encouraging and suggest that this new materials might be incorporated in future separation industrial processes.

REFERENCES

[1] Hernández, FJ; de los Ríos, AP; Rubio, M; Tomás–Alonso, F; Gómez, D & Víllora, G. A novel application of supported liquid membranes based on ionic liquids to the selective simultaneous separation of the substrates and products of a transesterification reaction, *J. Membr. Sci.*, 2007, 293, 73–80.

[2] Fortunato, R; González–Muñoz, MJ; Kubasiewicz, M; Luque, S; Alvarez, JR; Afonso, CAM; Coelhoso, IM & Crespo, JG. Liquid membranes using ionic liquids: The influence of water on solute transport, *J. Membr. Sci.*,2005, 249, 153–162.

[3] Kemperman, AJB; Bargeman, D; Boomgaard, T & Strathmann, H. Stability of supported liquid membranes: State of the art, *Sep. Sci. Technol.*, 1996, 31, 2733–2762.

[4] Takeuchi, H; Takahashi, K & Goto, G. Some observations on the stability of supported liquid membranes, *J. Membr. Sci.,* 1987, 34, 19–31.

[5] de los Rios, AP, Hernandez–Fernandez, FJ; Tomas–Alonso, F; Palacios, JM; Gomez, D; Rubio, M & Villora, G. A SEM–EDX study of highly stable supported liquid membranes based on ionic liquids, *J. Membr. Sci.,* 2007, 300, 88–94.

[6] Fortunato, R; Afonso, CAM; Benavente, J; Rodriguez–Castellon, E & Crespo, JG. Stability of supported ionic liquid membranes as studied by X–ray photoelectron spectroscopy, *J. Membr. Sci.*, 2005, 256, 216–223.

[7] Fortunato, R; Afonso, CAM; Reis, MAM & Crespo, JG. Supported liquid membranes using ionic liquids: study of stability and transport mechanism, *J. Membr. Sci.*, 2004, 242, 197–209.

[8] Brennecke, FJ & Maginn, EJ. Ionic liquids: innovative fluids for chemical processing, *AIChE J.,* 2001, 47, 2384–2389.

[9] Branco, LC; Grespo, JG & Afonso, CAM. Studies on the selective transport of organic compounds by using ionic liquids as a novel supported liquid membranes, *Chem. Eur. J.*, 2002, 8, 3865–3871.

[10] Noda, A; Hayamizu, K & Watanabe, M. Pulsed-Gradient Spin–Echo ^{1}H and ^{19}F NMR Ionic Diffusion Coefficient, Viscosity, and Ionic Conductivity of Non-Chloroaluminate Room-Temperature Ionic Liquids *J. Phys. Chem. B,* 2001, *105*, 4603-4610.

[11] Sun, J & Forsyth, M. Mac- Farlane, DR. Room temperature molten salts based on the quarternary ammonium ion, *J. Phys. Chem. B*. 1998, *102*, 8858.

[12] *Handbook of Chemistry & Physics*, 82nd Edition, D. R. Linde, Ed., CRC Press, New York, 2001, pp. 6-182 – 6-186.

[13] Bonhote, P; Dias, AP; Papageourgio, N; Kalayanasundaram, K & Gratzel, M. Hydrophobic, highly conductive ambient–temperature molten salts, *Inorg. Chem.*, 1996, 35, 1168–1178.

[14] Welton, T.Room–temperature ionic liquids. Solvents for synthesis and catalysis, *Chem. Rev.*, 1999, 99, 2071–2083.

[15] Seddon, K.Ionic liquids: designer solvents for green synthesis, *Chem. Eng.,* 2002, 730, 33–35.

[16] Branco, LC; Crespo, JG & Afonso, CAM. High selective transport of organic compounds by using supported liquid membranes based on ionic liquids, *Angew. Chem. Int. Ed.,* 2002, 41, 2771–2773.

[17] Matsumoto, M; Inomoto, Y & Kondo, K. Selective separation of aromatic hydrocarbons through supported liquid membranes based on ionic liquids, *J. Membr. Sci.,* 2005, 246, 77–81.

[18] Martak, J; Schlosser, S & Vlckova, S. Pertraction of lactic acid through supported liquid membranes containing phosphonium ionic liquid, *J. Membr. Sci.,* 2008, 318, 298–310.

[19] Scovazzo, P; Kieft, J; Finan, DA; Koval, C; DuBois, D & Noble, R. Gas separations using non–hexafluorophosphate anion supported ionic liquid membranes, *J. Membr. Sci.*, 2004, 238 57–63.

[20] de los Rios, AP; Hernandez–Fernandez, FJ; Rubio, M; Tomas–Alonso, F; Gomez, D & Víllora, G. Prediction of the selectivity in the recovery of transesterification reaction products using supported liquid membranes based on ionic liquids, *J. Membr. Sci.*, 2008, 307, 225–232.

[21] Ruiz, A; de los Rios, AP; Hernandez–Fernandez, FJ; Janssen, MHA; Schoevaart, R; van Rantwijk, F & Sheldon, RA. A cross–linked enzyme aggregate of *Candida antarctica* lipase B is active in denaturing ionic liquids, *Enzyme Microb. Technol.*, 2007, 40, 1095–1099.

[22] de los Rios, AP; Hernandez–Fernandez, FJ; Rubio, M; Gomez, D & Villora, G. Stabilization of native penicillin G acylase by ionic liquids, *J. Chem. Technol. Biotechnol.*, 2007, 82, 190–195.

[23] Feher, E; Illeova, V; Kelemen–Horvath, I; Belafi–Bako, K; Polakovic, M & Gubicza, L. Enzymatic production of isoamyl acetate in an ionic liquid–alcohol biphasic system, *J. Mol. Catal. B: Enzym.*, 2008, 50, 28–32.

[24] Frater, T; Gubicza, L; Szollösy, A & Bakos, J. Enantioselective hydrogenation in ionic liquids: Recyclability of the [Rh(COD)(DIPAMP)]BF4 catalyst in [bmim][BF_4], *Inorg. Chim. Acta*, 2006, 359, 2756–2759.

[25] Sheldon, R. Catalytic reactions in ionic liquids, *Chem. Commun.,* 2001, 23, 2399–2407.

[26] Earle, MJ & Seddon, KR. Ionic liquids. Green solvents for the future, *Pure Appl. Chem.*, 2000, 72, 1391–1398.

[27] Scott, K; Basov, N; Jachuck, RJJ; Winterton, N; Cooper, A & Davies, C. Reactor studies of supported ionic liquids rhodium–catalysed hydrogenation of propene, *Chem. Eng. Res. Des.*, 2005, 83, 1179–1185

[28] Hernández–Fernández, FJ; de los Ríos, AP; Tomás–Alonso, F; Gómez, D & Víllora, G. Preparation of supported ionic liquid membranes: Influence of the ionic liquid immobilization method on their operational stability, *J. Membr. Sci.,* 2009, 341, 172–177.

[29] Ropel, L; Belvèze, LS; Aki, SNVK; Stadtherr, MA & Brennecke, JF. Octanol–water partition coefficients of imidazolium–based ionic liquids, *Green Chem.*, 2005, **7**, 83–90.

[30] Gan, Q; Rooney, D; Xue, M; Thompson, G & Zou, Y. An experimental study of gas transport and separation properties of ionic liquids supported on nanofiltration membranes, *J. Membr. Sci.,* 2006, 280, 948–956.

[31] de los Ríos, AP; Hernández–Fernández, FJ; Tomás–Alonso, F; Palacios, JM & Víllora, G. Stability studies of supported liquid membranes based on ionic liquids: Effect of surrounding phase nature, *Desalination,* 2009, 245, 776–782.

[32] Bijani, S; Fortunato, R; Martínez de Yuso, MV; Heredia–Guerrero, FA; Rodríguez–Castellón, E; Coehloso, I; Crespo, J & Benavente, J. Physical–chemical and electrical characterizations of membranes modified with room temperature ionic liquids: Age effect, *Vacuum,* 2009, 83, 1283–1286.

[33] Cserjési, P; Nemestóthy, N; Vass, A; Csanádi, Zs & Bélafi–Bakó, K. Study on gas separation by supported liquid membranes applying novel ionic liquids, *Desalination*, 2009, 245, 743–747.

[34] Hanioka, S; Maruyama, T; Sotani, T; Teramoto, M; Matsuyama, H; Nakashima, K; Hanaki, M; Kubota, F & Goto, M. CO_2 separation facilitated by task–specific ionic liquids using a supported liquid membrane, *J. Membr. Sci.*, 2008, 314, 1–4.

[35] Scovazzo, P; Havard, D; McShea, M; Mixon, S & Morgan, D. Long–term, continuous mixed–gas dry fed CO2/CH4 and CO2/N2 separation performance and selectivities for room temperature ionic liquid membranes, *J. Membr. Sci.*, 2009, 327, 41–48.

[36] Ilconich, J; Myers, C; Pennline, H & Luebke, D. Experimental investigation of the permeability and selectivity of supported ionic liquid membranes for CO_2/He separation at temperatures up to 125 ºC, *J. Membr. Sci.*, 2007, 298, 41–47.

[37] Myers, C; Pennline, H; Luebke, D; Ilconichb, J; Dixon, JK; Maginn, EJ & Brennecke, JF. High temperature separation of carbon dioxide/hydrogen mixtures using facilitated supported ionic liquid membranes, *J. Membr. Sci.*, 2008, 322, 28–31.

[38] Hernández–Fernández, FJ; de los Ríos, AP; Tomás–Alonso, F; Gómez, D & Víllora, G. Kinetic resolution of 1–phenylethanol integrated with separation of substrates and products by a supported ionic liquid membrane, *J. Chem. Technol. Biotechnol,* 2007, 82, 190–195.

[39] de los Ríos, AP; Hernández–Fernández, FJ; Tomás–Alonso, F & Gómez, D. Víllora, Synthesis of flavour esters using free *Candida antarctica* lipase B in ionic liquids, *Flavour Fragrance J.*, 2008, 23, 319–322.

[40] Hernández–Fernández, FJ; de los Ríos, AP; Tomás–Alonso, F & Gómez, D. Víllora, Improvement in the separation efficiency of transesterification reaction compounds by the use of supported ionic liquid membranes based on the dicyanamide anion, *Desalination,* 2009, 244, 122–129.

[41] de los Ríos, AP; Hernández–Fernández, FJ; Rubio, M & Gómez, D. Víllora, Highly selective transport of transesterification reaction compounds through supported liquid membranes containing ionic liquids based on the tetrafluoroborate anion, *Desalination*, 2010, 250, 101–104.

[42] de los Ríos, AP; Hernández–Fernández, FJ; Presa, H & Gómez, D. Víllora, Tailoring supported ionic liquid membranes for the selective separation of transesterification reaction compounds, *J. Membr. Sci.,* 2009, 328, 81–85.

[43] Scovazzo, P, Determination of the upper limits, benchmarcks, and critical properties for gas separations using stabilized room temperature ionic liquid membranes (SILMs) for the purpose of guiding future research, *J. Membr. Sci.,* 2009, 343, 199–211.

[44] Bara, JE; Camper, DE & Gin, DL. Noble, RD, Room-Temperature Ionic Liquids and Composite Materials: Platform Technologies for CO2 Capture, *Accounts of Chemical Research,* 2010, 43, 152–159.

[45] Bara, JE; Carlisle, TK; Gabriel, JC, Camper, D; Finotello, A; Gin, DL & Noble, RD. Guide to CO2 Separations in Imidazolium-Based Room-Temperature Ionic Liquids, *Industrial and Engineering Chemistry Research*, 2009, 48, 2739–2751.

[46] Camper, D; Bara, J; Koval, C & Noble, R. Bulk-Fluid Solubility and Membrane Feasibility of Rmim-Based Room-Temperature Ionic Liquids, *Ind. Eng. Chem. Res.* 2006, *45,* 6279-6283.

[47] Izák, P; Köckerling, M & Kragl, U. Solute transport from aqueous mixture throught supported ionic liquid membrane by pervaporation, *Desalination*, 2006, 199, 96–98.

[48] Izák, P; Ruth, W; Fei, Z; Dyson, PJ & Kragl, U. Selective removal of acetone and butan–1–ol from water with supported ionic liquid–polydimethylsiloxane membrane by pervaporation, *Chem. Engin. J.,* 2008, 139, 318–321.

[49] Matsumoto, M; Ueba, K & Kondo, K. Vapor permeation of hydrocarbons through supported liquid membranes based on ionic liquids, *Desalination,* 2009, 241, 365–371.

[50] Peng, J; Liu, J; Hu, X & Jiang, G. Direct determination of chlorophenols in environmental water samples by hollow fiber supported ionic liquid membrane extraction coupled with high–performance liquid chromatography, *J. Chromatogr. A*, 2007, 1139, 165–170.

[51] Tao, Y; Liu, J; Hu, X; Li, H; Wang, T & Jiang, G. Hollow fiber supported ionic liquid membrane microextraction for determination of sulfonamides in environmental water samples by high–performance liquid chromatography, *J. Chromatogr. A*, 2009, 1216, 6259–6266.

In: Characterization and Development of Novel Materials ISBN: 978-1-61324-149-3
Editor: Rafiq Islam

Chapter 15

CRACKING STRESS AND BENDING MOMENT IN FLEXURAL FERROCEMENT ELEMENTS

***Zakaria Hossain*[1] *and Parvez Islam*[2*]**
[1]Faculty of Bioresources, Mie University, 1577 Kurima Machiya -cho, Tsu, Mie, Japan
[2]Faculty of Agricultural Engineering and Technology, Bangladesh Agricultural University, Bangladesh

ABSTRACT

The cracking stress and bending moment of ferrocement elements in flexure are the significant parameters in the design of structures made of ferrocement elements. This paper presents an experimental investigation on ferrocement elements uniformly reinforced by four different meshes, such as, geogrid mesh, chicken mesh, square mesh and hybrid mesh. The cracking stress and bending moment have been studied by testing the specimens under flexural loadings. By analyzing experimental data, the equations for the first-crack-stress and the ultimate-moment have been proposed in this paper. Results obtained by the proposed equations have been compared to those reported in the literature. It was revealed that the proposed equations, irrespective of their simplicity; give good agreement to those previously published.

Keywords: Ferrocement, Cracking stress, Bending Moment, Wire mesh, Flexural behavior

1. INTRODUCTION

Ferrocement has wider application in agriculture, housing industry, marine, water supply, sanitation, repair and strengthening of reinforced concrete structures in many developing and developed countries. These are usually subjected to transverse loads. Therefore, a thorough investigation on the flexural strength of ferrocement elements reinforced with different types

[*] Corresponding author: Email: engr_parvezfpm@yahoo.com.

of wire meshes is indispensable for its effective applications in construction A number of experimental and analytical studies have been conducted for the last two decades to establish the fundamental mechanical properties of ferrocement elements [1~28]. Among them Al-Rifaie and Aziz, and Ai-Rifaie and Trikha studied secondary strength of ferrocement with hexagonal mesh. Also, Hossain and Hasegawa [4,5] studied pre-post cracking behavior of ferrocement with square and hexagonal meshes. Cracking behavior and moment capacity of ferrocement reinforced with different reinforcements can be found in number of investigations [6~17]. In addition to these, tensile tests of ferrocement were conducted by Somayaji et al [18] and Velpari et al [19]. Some researches on ferrocement containing weldmesh and skeletal steel under flexural loading are also reported by Xion and Singh [20] and Yen and Su [21].

It is evident that in spite of the volume of information available, very little or no research work is reported in the literature on the comparison of various types of reinforcements in ferrocement as well as parameter identification of basic mechanical properties such as cracking stress and bending moments in flexure. The cracking stress and bending moment which represent the important structural properties in flexural design of ferrocement elements have not been received any attention yet, except for the research by Naaman and co-workers [22-24]. By drawing analogy with only one or two types of meshes in conventional ferrocement, one may argue that such mesh may behave similar to other types. However, this needs to be substantiated by evidence requiring experimentation and analytical modeling, if necessary. The present research program has been taken up to fulfill these basic needs.

Flexural tests are conducted on 48 ferrocement elements of sizes 400×200mm reinforced by four different types of wire meshes which are commonly used and are easily available in the local markets. They include chicken mesh, hybrid mesh, square mesh and geogrid mesh. Results of these tests are analyzed in detail to understand thoroughly the first crack stress, first crack moment and ultimate moment of the ferrocement elements. For the sake of generalized design convention, equations for calculation of first crack stress, first crack moment and ultimate moment are developed by analyzing the experimental data. The validity of the newly developed equations is established by comparing of the results obtained with them to those found in the literature.

Materials and Methods

Method of Specimen Preparation

The specimens are made using wooden moulds with open tops. The moulds are so made that each of the four side walls and the base of the form work are detachable so that the mould could be easily separated from the cast element after its initial setting. Four timber moulds are used in a batch. Indelible ink markings are made on all inside surfaces to indicate levels of different mesh layers during casting. A clear cover of 2 mm is maintained at the bottom and top sides. The type of mesh used and the number of layers of mesh are marked on the element. The meshes are uniformly arranged for all the specimens. The requisite amounts of sand and cement are mixed dry in a pan, and then the requisite quantity of water is added gradually while the mix is continuously stirred. The contact surfaces of the wooden mould to the mortar are greased before casting the specimens to ease the demoulding process.

Ordinary Portland cement and river sand passing through a No.8 (2.38mm) sieve, which has a fineness modulus of 2.33, are used for casting. For all the specimens, both the water-cement ratio and cement- sand ratio are maintained at 0.5 by weight. In each casting, two elements of plain mortar of size 400×200 mm with thickness 30 mm, and three cylinders of diameter 100 mm and length 120 mm are cast. These cylinders are used to test for determining the compressive strength, modulus of elasticity and Poisson's ratio of the mortar: these values are given in Table 1.

Table 1. Average values of plain mortar specimens

Specimens	Compressive Strength (N/mm^2)	Modulus of Elasticity (kN/mm^2)	Poisson's ratio
Rectangular	28.26	13.49	0.24
Cylindrical	27.43	17.45	0.25

Table 2. First crack stress, first crack moment and ultimate moment of ferrocement with geogrid mesh

Specimens	Rr (%)	Vr (%)	σ_{cr}, N/mm^2	Average σ_{cr} N/mm^2	M_{cr}, kN-m	Average Mcr, kN-m	M_{ult}, kN-m	Average M_{ult}, kN-m
G1	0.52	1.04	3.23	3.74	0.02	0.02	0.09	0.07
G1			4.25		0.02		0.05	
G2	1.03	2.03	3.98	4.07	0.04	0.05	0.07	0.10
G2			4.16		0.06		0.12	
G3	1.54	3.08	4.45	4.85	0.07	0.08	0.08	0.16
G3			5.25		0.09		0.24	
G4	2.06	4.12	6.15	5.71	0.11	0.13	0.17	0.24
G4			5.27		0.15		0.31	
G5	2.57	5.14	6.75	6.22	0.18	0.17	0.21	0.27
G5			5.69		0.16		0.32	
G6	3.08	6.16	6.35	6.86	0.2	0.22	0.27	0.28
G6			7.37		0.24		0.29	

Ordinary meshes obtained from the market are cut to obtain the requisite number of layers of desired size. The sand cement mortar layer is spread at the base of the mould, and on this base layer the first mesh is laid. The mesh layer is then covered by another layer of mortar, and the process is repeated until the specimen contains the desired number of mesh layers. Thus, each of the 30 mm-thick element has equally spaced mesh layers, leaving a cover of 2 mm at the top and bottom surfaces. The specimens are air-dried for 24 hours for initial setting and then immersed in water for curing. The specimens are removed from water after 28 days and were air-dried for 48 hours in room temperature of about 10o C and relative humidity of about 40%; then they are tested for their flexural performance.

Table 3. First crack stress, first crack moment and ultimate moment of ferrocement with chicken mesh

Specimens	R_r(%)	V_r(%)	σ_{cr} N/ mm^2	Average σ_{cr} N/ mm^2	M_{cr}, kN-m	Average M_{cr}, kN-m	M_{ult}, kN-m	Average M_{ult}, kN-m
C1	0.07	0.29	2.78	2.93	0.05	0.08	0.08	0.10
C1			3.08		0.11		0.12	
C2	0.14	0.59	4.48	4.12	0.19	0.17	0.23	0.19
C2			3.76		0.15		0.19	
C3	0.21	0.88	3.45	4.27	0.21	0.22	0.14	0.24
C3			5.09		0.23		0.34	
C4	0.28	1.17	5.86	5.29	0.34	0.26	0.41	0.30
C4			4.66		0.18		0.19	
C5	0.35	1.46	4.75	5.87	0.30	0.37	0.34	0.39
C5			6.99		0.44		0.44	
C6	0.42	1.76	6.43	6.83	0.32	0.41	0.4	0.43
C6			7.23		0.50		0.46	

Specification and Properties of Meshes

The geogrid is manufactured from polypropylene fiber. The junctions of the geogrid are directly connected and greatly improved by interweaving the yarns, and then the grids are coated with protective sheathing. The strength of the junctions is adequate to transmit the envisaged loadings. The cross-section of the geogrid strand is 3×4 mm size in both transverse and longitudinal directions with center to center openings of 22 mm. The Young's modulus and Poisson's ratio of the geogrid are reported as 75 kN/mm^2 and 0.4, respectively.

The diameters, Young's moduli and Poisson's ratios of the chicken mesh and square mesh, obtained from the open-market are 0.8 mm and 1.2 mm, 104 kN/mm^2 and 138 kN/mm^2, 0.3 and 0.3 respectively with c/c opening of square mesh is 10 mm. The hybrid mesh used in this research is the combination of chicken mesh and square mesh.

Method of Flexural Testing

All the elements are tested with their two edges simply supported over a span of 360 mm under two point-flexural loading. The distance between the two loading points is 120 mm with moment arms of 120 mm on both sides of the loading points. The tests are performed on a testing machine with 50 kN-capacity. The tests are performed with a loading speed of 1mm per minute and the deflection readings are taken at an interval of 0.1 kN. The deflections are measured with the dial gauges having a least count of 0.01 mm at the mid-section of the element. Before testing, all the elements are painted white so that cracks could be easily observed and clearly photographed. To be able to follow the development of the cracking pattern by visible observation, the tests are carried out with the tension side up. The load is applied in a vertically upward direction by means of an electric jack. The load application continued until the deflections became excessive. In general, in most of the elements initial cracks (visible to the naked eye) developed without any cracking noise. The crack patterns of some of the elements tested are shown in Figure 1.

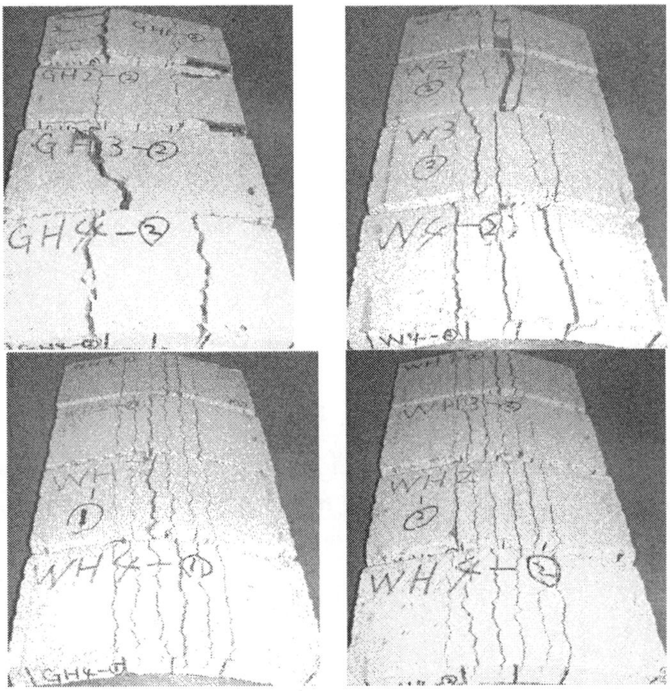

Figure 1. Crack pattern of some cement composites.

Effective Reinforcement

The effective reinforcement of the cement composite in a given direction is defined as the ratio of the area of mesh wires in that direction to the total area of specimen in the same direction. The percent effective reinforcement R_r for cement composite element containing square mesh in any direction after the cracking initiation can be written as;

$$R_r = \frac{25\pi d^2 N_L}{D t} \quad (1)$$

where, d is the diameter of wire, N_L is the number of layers, t is the thickness of the cement composite and D is the c/c distance of mesh wires. The numerical value 25 has mainly appeared due to calculation of effective reinforcement as a percentage. Equation (1) can be applied in case of square mesh reinforced specimens in which the mesh wires are parallel amongst each other. In case of chicken mesh, the wires are not parallel and hence the Eq. (1) gets modified as,

$$R_r = \frac{25\pi d^2 N_L Cos\,\theta}{D t} \tag{2}$$

where, θ is the angle of the mesh wires to the panel axis. θ has a value of 59.53 degrees for the chicken mesh used in the present research work.

Volume Fraction

The volume fraction V_r of ferrocement elements in any direction can be calculated by the following equation

$$V_r = \frac{25\pi d^2 N_L}{t}\left(\frac{1}{D_l} + \frac{1}{D_t}\right) \tag{3}$$

where, d , N_L and t have the same meaning as in Eqns. (1) and (2) while D_l and D_t are the c/c distance of mesh wires in the longitudinal and transverse directions, respectively. As in Eqns. 1 and 2, the numerical value 25 is visible due to conversion of volume fraction also in percentage basis.

Specific Surface

The specific surface of reinforcement in any direction can be defined as the ratio of the total surface area of reinforcement to the area of cross-section of ferrocement elements in that direction and can be expressed in the following form

$$S_{RL} = \frac{4V_r}{d_b} \tag{4}$$

where S_{RL} is the specific surface in mm^2/mm^3, V_r is volume fraction of reinforcement and d_b is the diameter of mesh wires in mm.

Cracking Stress of Ferrocement Element

Considering the homogeneous nature of the ferrocement composite, the first cracking stress (in flexure) can be obtained in the elastic range as

$$\sigma_{cr} = \frac{L t P_{cr}}{12 I} \tag{5}$$

where L = span length between the supports, t = thickness of the specimens, I = moment of inertia with respect to nutral axis and P_{cr} = first cracking load assumed very nearly equal to the maximum load.

RESULTS AND DISCUSSION

The first cracking stress, first cracking moment and ultimate moment of ferrocement elements reinforced with geogrid mesh, chicken mesh, square mesh and hybrid mesh in flexure obtained from the laboratory experiments are tabulated in Tables 2 to 5. These tables show that the first cracking stress, first cracking moment and ultimate moment of ferrocement elements vary with the variation of effective reinforcement (%) as well as with the variation of volume fraction (%) of reinforcement.

For the sake of clear understanding , first cracking stress of ferrocement elements with four different types of wire meshes, the data given in Tables 2 to 5 are plotted in Figure 2. It is evident that the first cracking stress increases with the increase in percent effective reinforcement for ferrocement containing any of the mesh-types. The rate of increase of first cracking stress, with effective reinforcement, is maximum for ferrocement elements containing the chicken mesh (volume fraction up to 1.76%) whereas it is minimum for ferrocement elements reinforced by geogrid mesh. It is also noted that the rate of increase of first cracking stress is more for hybrid mesh reinforced ferrocement than that of the square mesh reinforced ferrocement.

Table 4. First crack stress, first crack moment and ultimate moment of ferrocement with square mesh

Specimens	R_r(%)	V_r(%)	σ_{cr}, N/ mm^2	Average σ_{cr}, N/ mm^2	M_{cr}, kN-m	Average M_{cr}, kN-m	M_{ult}, kN-m	Average M_{ult}, kN-m
S1	0.38	0.76	6.10	5.55	0.60	0.49	0.7	0.65
S1			5.01		0.38		0.6	
S2	0.75	1.50	6.23	6.11	0.62	0.51	0.73	0.72
S2			5.98		0.40		0.71	
S3	1.13	2.26	6.45	7.04	0.49	0.54	0.54	0.83
S3			7.63		0.59		1.12	
S4	1.50	3.00	8.51	8.14	0.70	0.68	1	0.92
S4			7.78		0.66		0.84	
S5	1.69	3.38	9.25	8.56	0.75	0.72	1.1	0.95
S5			7.87		0.69		0.8	
S6	2.26	4.52	9.75	10.07	0.71	0.76	0.56	0.98
S6			10.39		0.81		1.4	

Table 5. First crack stress, first crack moment and ultimate moment of ferrocement with hybrid mesh

Specimens	R_r(%)	V_r(%)	σ_{cr}, N/ mm^2	Average σ_{cr}, N/ mm^2	M_{cr}, kN-m	Average M_{cr}, kN-m	M_{ult}, kN-m	Average M_{ult}, kN-m
H1	0.45	1.05	6.92	7.03	0.50	0.50	0.73	0.66
H1			7.14		0.50		0.59	
H2	0.52	1.35	7.82	7.61	0.55	0.53	0.79	0.75
H2			7.39		0.51		0.71	
H3	0.82	1.81	7.88	8.46	0.40	0.58	0.6	0.86
H3			9.05		0.76		1.12	
H4	0.89	2.11	9.77	9.50	0.71	0.66	1.3	0.93
H4			9.23		0.61		0.56	
H5	0.96	2.44	10.46	9.80	0.82	0.75	1.13	0.97
H5			9.15		0.68		0.81	
H6	1.34	3.14	11.97	11.50	0.90	0.82	1.4	1.10
H6			11.03		0.74		0.8	

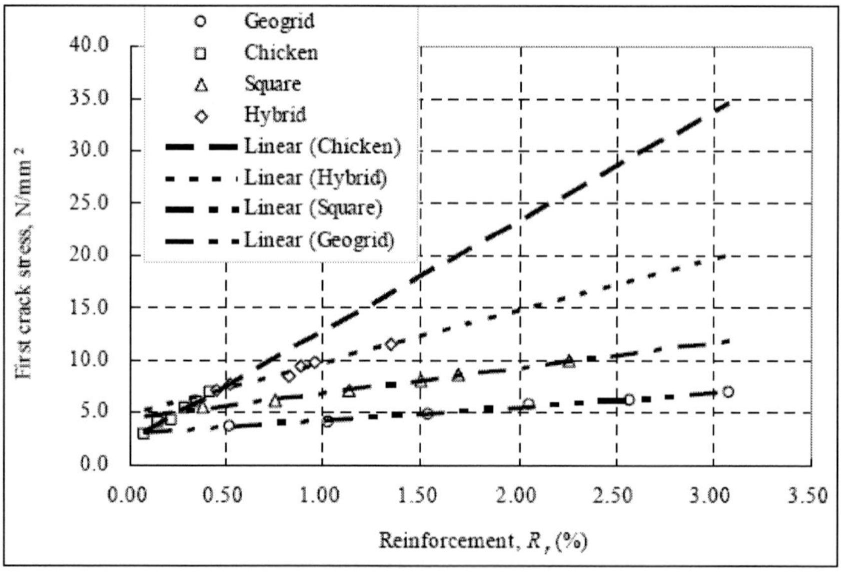

Figure 2. First crack stress vs. percent reinforcement.

For development of a generalized design equation for the first cracking stress for ferrocement elements reinforced by different types of meshes, the stress data given in Tables 2 to 5 are condensed in order to eliminate the effects of mesh type and the thickness of the ferrocement elements and the same is depicted in Figure 3. From the least square linearized curve of first cracking stress, drawn as a solid line in Figure 3, the following equation for the first cracking stress of ferrocement can be obtained.

$$\frac{\sigma_{cr}}{t} = \sigma_t + 0.094\, R_r \qquad (6)$$

where, σ_{cr} is the first cracking stress of the composite section in N/mm^2, σ_t is the tensile stress of the mortar per unit thickness, value calculated as 0.1246 N/mm^2/mm (intercept to y-axis) and R_r is the effective reinforcement in percent per unit thickness. It is noted here that the first cracking stress of ferrocement elements can be written as [22-24].

$$\sigma_{cr} = \sigma_{mu} + 25\, S_{RL} \qquad (7)$$

$$\sigma_{cr} = \sigma_{mu} + \sigma_R^{'}(V_r)^{1.1} \qquad (8)$$

where, σ_{mu} is the ultimate strength of mortar in N/mm^2, S_{RL} is the specific surface in mm^2/mm^3 (<0.2) in the loading direction, $\sigma_R^{'}$ is the proof stress of reinforcement corresponding to 0.01% strain (measured experimentally) and 1.1 is a empirical constant.

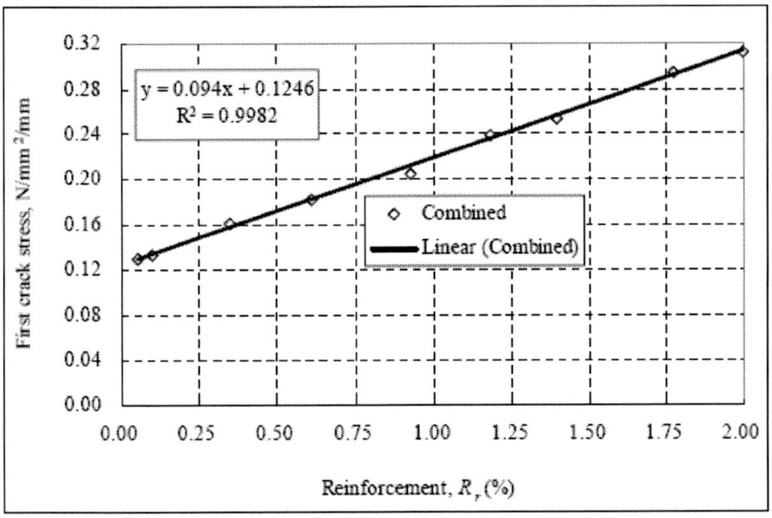

Figure 3. Condensed line of first crack stress vs. percent reinforcement.

The first crack moment and ultimate moment versus percent effective reinforcement of ferrocement elements reinforced by four types of wire meshes are given in Figures 4 and 5. Noticing from these figures, both the first crack moment and ultimate moment are increased with the increase in percent effective reinforcement. Similar to first crack stress, the rates of increment of the first crack moment and ultimate moment for ferrocement containing chicken meshes are highest whereas they are lowest for ferrocement containing geogrid meshes. Again the trend that was observed for the first crack stress is also found in the case of the first crack moment and ultimate moment in that both the moments at the first crack and ultimately are more for hybrid mesh than for geogrid and square meshes. From the curves given in Figures 2 - 5 , it is clearly implied that the ferrocement elements bear comparatively more loads when the mesh type changes from geogrid mesh to square mesh, from square mesh to hybrid mesh and from hybrid mesh to chicken mesh from the point of view of any of first crack stress, first crack moment or ultimate moment.

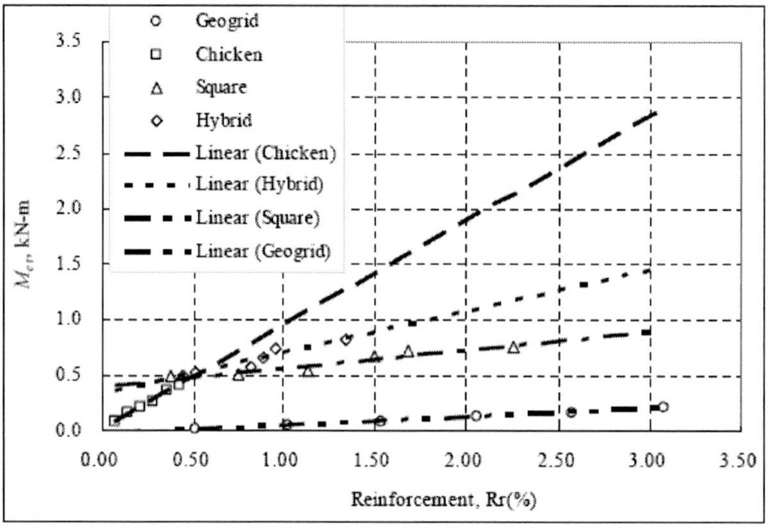

Figure 4. First crack moment vs. percent reinforcement.

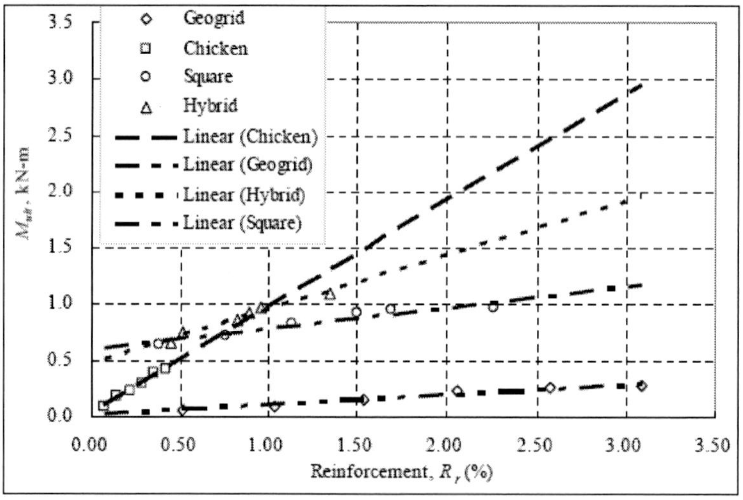

Figure 5. Ultimate moment vs. percent reinforcement.

For the sake of a clear perception and to eliminate the effect of thickness of ferrocement elements, the first crack moments and ultimate moments given in Figures 4 and 5 are condensed and are depicted in Figure 6. It is clear from this figure that the rate of increment of the ultimate moment containing any type of reinforcement, irrespective of thickness of ferrocement elements, is more than that the first crack moment of the ferrocement elements. From the linearized curves depicted in Figure 6, the following equations for the first crack moment and ultimate moment are obtained.

$$\frac{M_{cr}}{t} = M_t + 5.91 R_r \quad (9)$$

$$\frac{M_{ult}}{t} = M_t + 0.0515 + 10.556 R_r = M_t^{'} + 10.556 R_r \quad (10)$$

where, M_{cr} is the first crack moment of ferrocement elements with some reinforcements in kN-m, M_t is the moment at first crack for the section with zero percent reinforcement, value calculated as 7.4925 kN-m/m (intercept to y-axis) and R_r is the effective reinforcement in percent per unit thickness and M_{ult} is the ultimate moment of ferrocement elements with some reinforcements in kN-m, $M_t^{'}$ is the moment at ultimate load for the section with zero percent reinforcement, value calculated as 7.544kN-m/m (intercept to y-axis) The value of M_t and $M_t^{'}$ should be the same for the composite section of zero percent reinforcement i.e. the linearized curves of the first crack moment and the ultimate moment should coincide with each other at the zero percent reinforcement because plain mortar specimens (zero reinforcement composite) can never support any moment beyond the first crack moment. However, this difference is not so high and thus has no significant effect in calculating the ultimate moment because ferrocement always contains some reinforcement which forces the calculation to be beyond the neighborhood of zero reinforcement and consequently, the discrepancy becomes insignificant for the ferrocement elements.

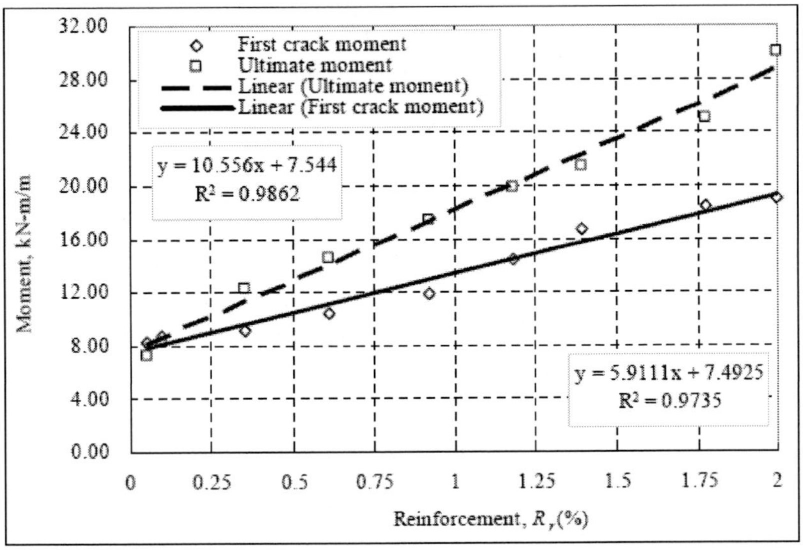

Figure 6. Condensed lines of moments vs. percent reinforcement.

It is noted here that the R-square or the coefficient of determination of the regression analysis has the values of 0.9982, 0.9735 and 0.9862 for the ferrocement elements at first crack stress, first crack moment and ultimate moment respectively containing any type of reinforcement i.e. the R-square value for all the cases close to 1.0 indicates that the test-data are fitted well and we have accounted for almost all of the variability with the variables specified in this paper.

It has been shown earlier that the ultimate moment of ferrocement element in flexure can be calculated by the following equation [27].

$$\frac{M_{ult}}{f_c' b t^2 \eta} = 0.0049 + 0.42 V_r \frac{f_y}{f_c'} - 0.077 \left(V_r \frac{f_y}{f_c'}\right)^2 \qquad (11)$$

where, M_{ult} is the ultimate flexural moment of ferrocement element, f_c' is the mortar compressive strength, b is the width of the specimens, t is the thickness, η is the global efficiency factor of mesh in the loading direction, f_y is the yield strength of reinforcement and V_r is the volume fraction of reinforcement.

Table 6. First crack stress of ferrocement elements- A comparison

R_r(%)	V_r(%)	S_{RL}	σ_{cr},(Eq.6) N/ mm^2	σ_{cr},(Eq.12) N/ mm^2	σ_{cr},(Eq.7) N/ mm^2	σ_{cr},(Eq.8) N/ mm^2
1.00	2.00	0.08	6.42	13.39	2.67	6.08
1.50	3.00	0.12	7.83	16.34	3.67	9.12
2.00	4.00	0.16	9.24	19.29	4.67	12.27
2.50	5.00	0.20	10.65	22.24	5.67	15.49
3.00	6.00	0.24	12.06	25.19	6.67	18.78
3.50	7.00	0.28	13.47	28.14	7.67	22.13
4.00	8.00	0.32	14.88	31.09	8.67	25.53

Table 7. Ultimate moment of ferrocement elements - A comparison

R_r(%)	V_r(%)	M_{ult} (Eq.10) N-m	M_{ult} (Eq.11) N-m	Eq.10/Eq.11
1.00	2.00	541.50	331.09	1.64
1.50	3.00	699.75	475.50	1.47
2.00	4.00	858.00	609.91	1.41
2.50	5.00	1016.25	734.32	1.38
3.00	6.00	1174.50	848.73	1.38
3.50	7.00	1332.75	953.14	1.40
4.00	8.00	1491.00	1047.56	1.42

The first crack stress calculated by the proposed Eqs. 6 and 9 are compared with those obtained by previously published Eqs. 7 and 8 and are depicted in Table 6. It is noted here that the following relationship is used in calculating the first crack stress from the first crack moment of the ferrocement elements.

$$\sigma_{cr} = \frac{M_{cr}t}{2I} \qquad (12)$$

where, σ_{cr} is the first crack stress, M_{cr} is the first crack moment, t is the thickness of the ferrocement elements and I is moment of inertia. Also the ultimate moments obtained by the proposed Eq.(10) here are compared with the earlier reported Eq.(11) and the results are depicted in Table 7. It is obvious that both the first crack stress and ultimate moment obtained by the presently proposed equations are in reasonably good agreement with the results of earlier equations, even though the proposed equations are very simple.

CONCLUSION

1. Flexural first crack stress, first crack moment and ultimate moment increased with the increase in percent effective reinforcement for any type of wire meshes.
2. The rate of increase of both the cracking stress and the ultimate bending moment are maximum for ferrocement containing chicken mesh and are the least for the geogrid mesh.
3. The proposed equations for the first crack stress, first crack moment and ultimate moment of flexural ferrocement elements are simple but provide reasonably accurate results as compared to relatively more complicated equations involving many parameters.

REFERENCES

[1] ACI Committee 549, "Guide for Design, Construction, and Repair of ferrocement", *ACI Structural Journal*, May-June, pp.325-351, 1988.

[2] Al-Rifaie, W. N. and Aziz, A. A. "Experimental Investigation of Secondary Strength of Ferrocement Reinforced with hexagonal Mesh", *J. of Ferrocement,* 17(3), pp.215-222, 1987.

[3] Al-Rifaie, W. N. and Trikha, D. N., "Effect of arrangement and orientation of hexagonal mesh on the behavior of two-way ferrocement slabs", *J. of Ferrocement*, Vol.20, No.3, pp.219-230, 1990.

[4] Hossain M. Z. and Hasegawa T., "A Comparison of the Mechanical Properties of Ferrocement in Flexure for Square and Hex Meshes", *Journal of Ferrocement*, Vol.28, No.2, pp.11-134, 1998.

[5] Hossain, M. Z. and Hasegawa T., "A Study on Pre-and Post-Cracking Behavior of Ferrocement Plates" *J. of Ferrocement*, Vol.27, No.2, pp.127-143, 1997.

[6] Hussin, M. W., "Deflection and cracking performance of fibrous ferrocement thin sheets" *J. of Ferrocement*, Vol.21, No.1, pp.31-41, 1991.

[7] Johnston, C. D. and Mowat, D. N., "Ferrocement - Material Behavior in Flexure" *J. of Structural Division*, ASCE, 100(10), pp.2053-2069, 1974.

[8] Kaushik, S. K., Singh, K. K. and Prasad, R., “Buckling of ferrocement plates”, *J. of Ferrocement*, Vol.24, No.1, pp.7-16, 1994.

[9] Kobayashi, Y., Tanaka, Y. and Ono, M., “Flexural impact damage of ferrocement”, *J. of Ferrocement*, Vol.22, No.3, pp.249-263, 1992.

[10] Logan, D. and Shah, S. P., “Moment Capacity and Cracking Behavior of Ferrocement in Flexure”, *ACI Journal,* 70(12), pp.799-804, 1973.

[11] Mansur, M. A. and Paramasivam, P., “Cracking behavior and ultimate strength of ferrocement in flexure”, *J. of Ferrocement*, Vol.16, No.4, pp.405-416, 1986.

[12] Natesan, S. C. and Rajasekaran, S., “Cracking and leakage of fiber reinforced ferrocement tanks, an experimental study”, *J. of Ferrocement*, Vol.14, No.1, pp.1-10, 1984.

[13] Onet, T., Magureanu, C and Vescan, V., “Aspect concerning the behavior of ferrocement in flexure”, *J. of Ferrocement*, Vol.22, No.1, pp.1-9, 1992.

[14] Ong, K. C. G., Paramasivam, P. and Lim, C. T. T., “Flexural strengthening of reinforced concrete beams using ferrocement laminates”, *J. of Ferrocement*, Vol.22, No.4, pp.331-342, 1992.

[15] Rosenthal, I. and Bljuger, F., “Bending behavior of ferrocement-reinforced concrete composite beams”, *J. of Ferrocement*, Vol.15, No.1, pp.15-24, 1985.

[16] Shirai, A. and Ohama, Y. “Improvement in flexural behavior and impact resistance of ferrocement by use of polymers”, *J. of Ferrocement*, Vol.20, No.3, pp.257-264, 1990.

[17] Singh, G. et.al. “Effect of Repeated Loading on Crack Width of Ferrocement, *J. of Ferrocement”*, vol.21, No.2, pp.119-126, 1991.

[18] Somayaji, S. and Naaman, A. E., “Stress-strain response and cracking of ferrocement in tension”, *J. of Ferrocement*, Vol.11, No.2, pp.127-142, 1981.

[19] Velpari, V., Raghavendra, N.M. and Balasubramanian, N., “Tensile and flexural properties of polymer impregnated ferrocement composites”, *J. of Ferrocement*, Vol.11, No.2, pp.143-154, 1981.

[20] Xiong, G. J. and Singh, G., “Behavior of weldmesh ferrocement composite under flexural cyclic loading”, *J. of Ferrocement*, Vol.23, No.3, pp.237-238, 1992.

[21] Yen, T. and Su, C. F., “The influence of skeletal steel on the flexural behavior of ferrocement”, *J. of Ferrocement*, Vol.10, No.3, pp.177-188, 1980.

[22] Naaman, A. E. and Shah, S. P. Tensile Tests of Ferrocement, *ACI Journal*, 68(9), pp.693-698, 1971.

[23] Naaman, A. E. and Homrich, J. R., “Flexural design of ferrocement: Computerized evaluation and design aids”, *J. of Ferrocement*, Vol.16, No.2, pp.101-116, 1986.

[24] Naaman, A. E., “Ferrocement and Laminated Cementitious Composites”, *Techno Press 3000, Ann Arbor, Michigan*, 2000.

INDEX

C

D

G

H

I

N

O

P

Q

R

S

T

U

V

W

X

Y

Z